Courtright Memorial Library
Otterbein University
138 W. Main St.
Westerville, Ohio 43081

NUMERICAL MATHEMATICS AND SCIENTIFIC COMPUTATION

Series Editors
A.M. STUART   E. SÜLI

NUMERICAL MATHEMATICS AND SCIENTIFIC COMPUTATION

Books in the series
Monographs marked with an asterix (*) appeared in the series 'Monographs in Numerical Analysis' which has been folded into, and is continued by, the current series.

For a full list of titles please visit
http://www.oup.co.uk/academic/science/maths/series/nmsc

* J. H. Wilkinson: *The algebraic eigenvalue problem*
* I. Duff, A.Erisman, and J. Reid: *Direct methods for sparse matrices*
* M. J. Baines: *Moving finite elements*
* J.D. Pryce: *Numerical solution of Sturm–Liouville problems*

Ch. Schwab: *p- and hp- finite element methods: theory and applications to solid and fluid mechanics*
J.W. Jerome: *Modelling and computation for applications in mathematics, science, and engineering*
Alfio Quarteroni and Alberto Valli: *Domain decomposition methods for partial differential equations*
G.E. Karniadakis and S.J. Sherwin: *Spectral/hp element methods for CFD*
I. Babuška and T. Strouboulis: *The finite element method and its reliability*
B. Mohammadi and O. Pironneau: *Applied shape optimization for fluids*
S. Succi: *The Lattice Boltzmann Equation for fluid dynamics and beyond*
P. Monk: *Finite element methods for Maxwell's equations*
A. Bellen & M. Zennaro: *Numerical methods for delay differential equations*
J. Modersitzki: *Numerical methods for image registration*
M. Feistauer, J. Felcman, and I. Straškraba: *Mathematical and computational methods for compressible flow*
W. Gautschi: *Orthogonal polynomials: computation and approximation*
M.K. Ng: *Iterative methods for Toeplitz systems*
Michael Metcalf, John Reid, and Malcolm Cohen: *Fortran 95/2003 explained*
George Em Karniadakis and Spencer Sherwin: *Spectral/hp element methods for CFD, second edition*
Dario A. Bini, Guy Latouche, and Beatrice Meini: *Numerical methods for structured Markov chains*
Howard Elman, David Silvester, and Andy Wathen: *Finite elements and fast iterative solvers: with applications in incompressible fluid dynamics*
Moody Chu and Gene Golub: *Inverse eigenvalue problems: Theory and applications*
Jean-Frédéric Gerbeau, Claude Le Bris, and Tony Lelièvre: *Mathematical methods for the magnetohydrodynamics of liquid metals*
Grégoire Allaire: *Numerical analysis and optimization*
Eric Cancès, Claude Le Bris, Yvon Maday, and Gabriel Turinici: *An introduction to mathematical modelling and numerical simulation*
Karsten Urban: *Wavelet methods for elliptic partial differential equations*
B. Mohammadi and O. Pironneau: *Applied shape optimization for fluids, second edition*

APPLIED SHAPE OPTIMIZATION FOR FLUIDS
2nd Edition

Bijan Mohammadi
University Montpellier II
Olivier Pironneau
University Paris VI

OXFORD
UNIVERSITY PRESS

Great Clarendon Street, Oxford OX2 6DP

Oxford University Press is a department of the University of Oxford.
It furthers the University's objective of excellence in research, scholarship,
and education by publishing worldwide in

Oxford New York

Auckland Cape Town Dar es Salaam Hong Kong Karachi
Kuala Lumpur Madrid Melbourne Mexico City Nairobi New Delhi
Shanghai Taipei Toronto

With offices in

Argentina Austria Brazil Chile Czech Republic France Greece
Guatemala Hungary Italy Japan Poland Portugal
Singapore South Korea Switzerland Thailand Turkey Ukraine Vietnam

Oxford is a registered trade mark of Oxford University Press
in the UK and in certain other countries

Published in the United States
by Oxford University Press Inc., New York

© Oxford University Press 2010

The moral rights of the authors have been asserted
Database right Oxford University Press (maker)

First published 2010

All rights reserved. No part of this publication may be reproduced,
stored in a retrieval system, or transmitted, in any form or by any means,
without the prior permission in writing of Oxford University Press,
or as expressly permitted by law, or under terms agreed with the appropriate
reprographics rights organization. Enquiries concerning reproduction
outside the scope of the above should be sent to the Rights Department,
Oxford University Press, at the address above

You must not circulate this book in any other binding or cover
and you must impose this same condition on any acquirer

British library catalogue in Publication Data

Data available

Library of Congress Cataloging-in-Publication Data

Data available

Typeset by Author using LaTex
Printed in Great Britain
on acid-free paper by the
MPG Books Group, Bodmin and King's Lynn

ISBN 978–0–19–954690–9

1 3 5 7 9 10 8 6 4 2

We dedicate this second edition to our late master Jacques-Louis Lions.

Professor J.-L. Lions passed away in 2001; at the time this book was written he was also the chief scientific advisor to the CEO at Dassault-aviation; our gratitude goes to him for his renewed encouragement and support.

PREFACE

The first edition of this book was written in 2001 when computers in industry were hardly sufficient to optimize shapes for fluid problems. Since then computers have increased twenty fold in power; consequently methods which were not feasible have begun giving results, namely evolutionary algorithms, topological optimization methods and level set algorithms. While these were mentioned briefly in the first edition, here they now have separate chapters. Yet the book remains mostly a differential shape optimization book and our coverage of these three new methods is still minimal, each requiring in fact a separate book. To our credit, it should also be said that genetic algorithms are not yet capable of solving problems like wing optimization when the number of parameters is bigger than a few dozen without intensive distributed resources; similarly topological optimization is great for structure optimization but only an interesting alternative for fluid flows in most cases. Level sets, on the other hand, are more general but simply another parameterization method; the optimization is done with a gradient or Newton algorithm, so it is within the scope of the book.

ACKNOWLEDGEMENTS

The authors are grateful to F. Alauzet, R. Arina, P. Aubert, A. Baron, R. Brahadwaj, L. Debiane, N. Dicesaré, F. Hecht, S. Galera, B. Ivorra, D. Isèbe, G. Medic, N. Petruzzelli, G. Puigt, J. Santiago, M. Stanciu, E. Polak and J. Tuomela for their contributions in the form of scientific material published elsewhere in collaboration with us.

For their encouragement and sharing of ideas the authors would like to thank A. Dervieux, H.-G. Bock, C. Farhat, M. Giles, R. Glowinski, M. Gunzburger, W. Habashi, M. Hafez, A. Henrot, D. Hertzog, H. Kawarada, P. Le Tallec, P. Moin, M. Navon, P. Neittanmaaki, J. Periaux, B. Perthame, P. Sagaut, S. Obayashi, M. Wang.

We thank also our colleagues at the universities of Montpellier II and Paris VI and at INRIA, for their comments on different points related to this work, namely: H. Attouch, P. Azerad, F. Bouchette, M. O. Bristeau, J. F. Bourgat, M. Cuer, A. Desideri, P. Frey, A. Hassim, P.L. George, B. Koobus, S. Lanteri, P. Laug, E. Laporte, F. Marche, A. Marrocco, F. Nicoud, P. Redont, E. Saltel, M. Vidrascu.

We are also very happy to acknowledge the contributions of our industrial partners: MM. Duffa, Pirotais, Galais, Canton-Desmeuzes, at CEA-CESTA; MM. Stoufflet, Mallet, Rostand, Rogé, Dinh at Dassault Aviation, MM. Chaput, Cormery and Meaux at Airbus, MM. Chabard, Laurence and Viollet at EDF. MM. Aupoix, Cousteix at Onera. S. Moreau at Valeo. MM. Poinsot and André at Cerfacs.

Finally, considerable help was given to us by the automatic differentiation specialists and especially by C. Bishof, C. Faure, P. Hovland, N. Rostaing, A. Griewank, J.C. Gilbert and L. Hascoet.

As this list is certainly incomplete, many thanks and our apologies to colleagues whose name is missing.

CONTENTS

1 **Introduction**   1

2 **Optimal shape design**   6
   2.1 Introduction   6
   2.2 Examples   7
     2.2.1 Minimum weight of structures   7
     2.2.2 Wing drag optimization   8
     2.2.3 Synthetic jets and riblets   11
     2.2.4 Stealth wings   12
     2.2.5 Optimal breakwater   15
     2.2.6 Two academic test cases: nozzle optimization   16
   2.3 Existence of solutions   17
     2.3.1 Topological optimization   17
     2.3.2 Sufficient conditions for existence   18
   2.4 Solution by optimization methods   19
     2.4.1 Gradient methods   19
     2.4.2 Newton methods   20
     2.4.3 Constraints   21
     2.4.4 A constrained optimization algorithm   22
   2.5 Sensitivity analysis   22
     2.5.1 Sensitivity analysis for the nozzle problem   25
     2.5.2 Numerical tests with freefem++   27
   2.6 Discretization with triangular elements   28
     2.6.1 Sensitivity of the discrete problem   30
   2.7 Implementation and numerical issues   33
     2.7.1 Independence from the cost function   33
     2.7.2 Addition of geometrical constraints   34
     2.7.3 Automatic differentiation   34
   2.8 Optimal design for Navier-Stokes flows   35
     2.8.1 Optimal shape design for Stokes flows   35
     2.8.2 Optimal shape design for Navier-Stokes flows   36
   References   37

3 **Partial differential equations for fluids**   41
   3.1 Introduction   41
   3.2 The Navier-Stokes equations   41
     3.2.1 Conservation of mass   41
     3.2.2 Conservation of momentum   41
     3.2.3 Conservation of energy and and the law of state   42
   3.3 Inviscid flows   43

3.4	Incompressible flows	44
3.5	Potential flows	44
3.6	Turbulence modeling	46
	3.6.1 The Reynolds number	46
	3.6.2 Reynolds equations	46
	3.6.3 The $k - \varepsilon$ model	47
3.7	Equations for compressible flows in conservation form	48
	3.7.1 Boundary and initial conditions	50
3.8	Wall laws	51
	3.8.1 Generalized wall functions for u	51
	3.8.2 Wall function for the temperature	53
	3.8.3 k and ε	54
3.9	Generalization of wall functions	54
	3.9.1 Pressure correction	54
	3.9.2 Corrections on adiabatic walls for compressible flows	55
	3.9.3 Prescribing ρ_w	56
	3.9.4 Correction for the Reichardt law	57
3.10	Wall functions for isothermal walls	58
	References	60

4 Some numerical methods for fluids — 61

4.1	Introduction	61
4.2	Numerical methods for compressible flows	61
	4.2.1 Flux schemes and upwinded schemes	61
	4.2.2 A FEM-FVM discretization	62
	4.2.3 Approximation of the convection fluxes	63
	4.2.4 Accuracy improvement	64
	4.2.5 Positivity	64
	4.2.6 Time integration	65
	4.2.7 Local time stepping procedure	66
	4.2.8 Implementation of the boundary conditions	66
	4.2.9 Solid walls: transpiration boundary condition	67
	4.2.10 Solid walls: implementation of wall laws	67
4.3	Incompressible flows	68
	4.3.1 Solution by a projection scheme	69
	4.3.2 Spatial discretization	70
	4.3.3 Local time stepping	71
	4.3.4 Numerical approximations for the $k - \varepsilon$ equations	71
4.4	Mesh adaptation	72
	4.4.1 Delaunay mesh generator	72
	4.4.2 Metric definition	73
	4.4.3 Mesh adaptation for unsteady flows	75
4.5	An example of adaptive unsteady flow calculation	77
	References	78

5	Sensitivity evaluation and automatic differentiation	81
	5.1 Introduction	81
	5.2 Computations of derivatives	83
	5.2.1 Finite differences	83
	5.2.2 Complex variables method	83
	5.2.3 State equation linearization	84
	5.2.4 Adjoint method	84
	5.2.5 Adjoint method and Lagrange multipliers	85
	5.2.6 Automatic differentiation	86
	5.2.7 A class library for the direct mode	88
	5.3 Nonlinear PDE and AD	92
	5.4 A simple inverse problem	94
	5.5 Sensitivity in the presence of shocks	101
	5.6 A shock problem solved by AD	103
	5.7 Adjoint variable and mesh adaptation	104
	5.8 `Tapenade`	106
	5.9 Direct and reverse modes of AD	106
	5.10 More on `FAD` classes	109
	References	113
6	Parameterization and implementation issues	116
	6.1 Introduction	116
	6.2 Shape parameterization and deformation	116
	6.2.1 Deformation parameterization	117
	6.2.2 CAD-based	117
	6.2.3 Based on a set of reference shapes	117
	6.2.4 CAD-free	118
	6.2.5 Level set	122
	6.3 Handling domain deformations	127
	6.3.1 Explicit deformation	128
	6.3.2 Adding an elliptic system	129
	6.3.3 Transpiration boundary condition	129
	6.3.4 Geometrical constraints	131
	6.4 Mesh adaption	133
	6.5 Fluide-structure coupling	136
	References	138
7	Local and global optimization	140
	7.1 Introduction	140
	7.2 Dynamical systems	140
	7.2.1 Examples of local search algorithms	140
	7.3 Global optimization	142
	7.3.1 Recursive minimization algorithm	143
	7.3.2 Coupling dynamical systems and distributed computing	144

	7.4	Multi-objective optimization	145
		7.4.1 Data mining for multi-objective optimization	148
	7.5	Link with genetic algorithms	150
	7.6	Reduced-order modeling and learning	153
		7.6.1 Data interpolation	154
	7.7	Optimal transport and shape optimization	158
	References		161
8	**Incomplete sensitivities**		**164**
	8.1	Introduction	164
	8.2	Efficiency with AD	165
		8.2.1 Limitations when using AD	165
		8.2.2 Storage strategies	166
		8.2.3 Key points when using AD	167
	8.3	Incomplete sensitivity	168
		8.3.1 Equivalent boundary condition	168
		8.3.2 Examples with linear state equations	169
		8.3.3 Geometric pressure estimation	171
		8.3.4 Wall functions	172
		8.3.5 Multi-level construction	172
		8.3.6 Reduced order models and incomplete sensitivities	173
		8.3.7 Redefinition of cost functions	174
		8.3.8 Multi-criteria problems	175
		8.3.9 Incomplete sensitivities and the Hessian	175
	8.4	Time-dependent flows	176
		8.4.1 Model problem	178
		8.4.2 Data mining and adjoint calculation	181
	References		183
9	**Consistent approximations and approximate gradients**		**184**
	9.1	Introduction	184
	9.2	Generalities	184
	9.3	Consistent approximations	186
		9.3.1 Consistent approximation	187
		9.3.2 Algorithm: conceptual	187
	9.4	Application to a control problem	188
		9.4.1 Algorithm: control with mesh refinement	189
		9.4.2 Verification of the hypothesis	189
		9.4.3 Numerical example	190
	9.5	Application to optimal shape design	190
		9.5.1 Problem statement	191
		9.5.2 Discretization	192
		9.5.3 Optimality conditions: the continuous case	192
		9.5.4 Optimality conditions: the discrete case	193
		9.5.5 Definition of θ_h	194

		9.5.6 Implementation trick	195
		9.5.7 Algorithm: OSD with mesh refinement	195
		9.5.8 Orientation	196
		9.5.9 Numerical example	196
		9.5.10 A nozzle optimization	197
		9.5.11 Theorem	199
		9.5.12 Numerical results	200
		9.5.13 Drag reduction for an airfoil with mesh adaptation	200
	9.6	Approximate gradients	203
		9.6.1 A control problem with domain decomposition	204
		9.6.2 Algorithm	205
		9.6.3 Numerical results	207
	9.7	Conclusion	209
	9.8	Hypotheses in Theorem 9.3.2.1	209
		9.8.1 Inclusion	209
		9.8.2 Continuity	209
		9.8.3 Consistency	209
		9.8.4 Continuity of θ	209
		9.8.5 Continuity of $\theta_h(\alpha_h)$	210
		9.8.6 Convergence	210
	References		210
10	**Numerical results on shape optimization**		**212**
	10.1	Introduction	212
	10.2	External flows around airfoils	213
	10.3	Four-element airfoil optimization	213
	10.4	Sonic boom reduction	215
	10.5	Turbomachines	217
		10.5.1 Axial blades	219
		10.5.2 Radial blades	222
	10.6	Business jet: impact of state evaluations	225
	References		225
11	**Control of unsteady flows**		**227**
	11.1	Introduction	227
	11.2	A model problem for passive noise reduction	228
	11.3	Control of aerodynamic instabilities around rigid bodies	229
	11.4	Control in multi-disciplinary context	229
		11.4.1 A model problem	230
		11.4.2 Coupling strategies	236
		11.4.3 Low-complexity structure models	237
	11.5	Stability, robustness, and unsteadiness	241
	11.6	Control of aeroelastic instabilities	244
	References		245

12 From airplane design to microfluidics — 246
- 12.1 Introduction — 246
- 12.2 Governing equations for microfluids — 247
- 12.3 Stacking — 247
- 12.4 Control of the extraction of infinitesimal quantities — 249
- 12.5 Design of microfluidic channels — 249
  - 12.5.1 Reduced models for the flow — 255
- 12.6 Microfluidic mixing device for protein folding — 255
- 12.7 Flow equations for microfluids — 259
  - 12.7.1 Coupling algorithm — 260
- References — 261

13 Topological optimization for fluids — 263
- 13.1 Introduction — 263
- 13.2 Dirichlet conditions on a shrinking hole — 264
  - 13.2.1 An example in dimension 2 — 264
- 13.3 Solution by penalty — 265
  - 13.3.1 A semi-analytical example — 267
- 13.4 Topological derivatives for fluids — 268
  - 13.4.1 Application — 268
- 13.5 Perspective — 270
- References — 270

14 Conclusions and prospectives — 272

Index — 275

1
INTRODUCTION

Nowadays the art of computer simulation has reached some maturity; and even for still unsolved problems engineers have learned to extract meaningful answers and trends for their design from rough simulations: numerical simulation is one of the tools on which intuition can rely! Yet for those who want to study trends and sensitivities more rationally the tools of automatic differentiation and optimization are there. This book deals with them and their application to the design of the systems of fluid mechanics. But brute force optimization is too often an inefficient approach and so our goal is not only to recall some of the tools but also to show how they can be used with some subtlety in an optimal design program.

Optimal shape design (OSD) is now a necessity in several industries. In airplane design, because even a few percent of drag reduction means a lot, aerodynamic optimization of 3D wings and even wing body configurations is routinely done in the aeronautics industry. Applications to the car industry are well underway especially for the optimization of structures to reduce weight but also to improve vehicle aerodynamics. Optimization of pipes, heart valves, and even MEMS and fluidic devices, is also done. In electromagnetism stealth objects and antenna are optimized subject to aerodynamic constraints.

However, OSD is still a difficult and computer-intensive task. Several challenges remain. One is multi-objective design. In aeronautics, high lift configurations are also challenging because the flow needs to be accurately solved and turbulence modelling using DES or LES is still too heavy to be included in the design loop, but also because shape optimization for unsteady flows is still immature.

From a mathematical point of view, OSD is also difficult because even if the problem is well posed success is not guaranteed. One should pay attention to the computing complexity and use sub-optimal approaches whenever possible. As we have said, demand is on multi-disciplinary and multi-criteria design and local minima are often present; a good treatment of state constraints is also a numerical challenge. Global optimization approaches based on a mix of deterministic and nondeterministic methods, together with surface response model reduction, is necessary to break complexity. Care should also be taken when noise is present in the data and always consider robustness issues.

From a theoretical point of view, OSD problems can be studied as infinite dimensional controls with state variables in partial differential equations and constraints. The existence of a solution is guaranteed under mild hypothesis in 2D and under the flat cone property in 3D. Tikhonov regularization is easily done with penalization of the surface of the shape. In variational form results translate without modifications to the discrete cases if discretized by the finite element or finite volume methods. Gradient methods are efficient and convergent even though it is always preferable to use second order methods when possible. Geometric constraints can be handled at no cost but more complex constraints involving the state variables are a real challenge. Multicriteria optimization and Pareto optimality have not been solved in a satisfactory way by differentiable optimization, either because the problems are too stiff and/or there are too many local minima. Evolutionary algorithms offer an expensive alternative. The black box aspect of this solution is a real asset in the industrial context. The consensus seems to go to a mix of stochastic and deterministic approaches using reduced order or surrogate models when possible. Topological optimization is a very powerful tool for optimizing the coefficients of PDEs. It is ideal for structure optimization where the answer can be a composite material or for low Reynolds flows. However, it does not look to be a promising technique for high Reynolds number flow.

Different choices can be made for the shape parameter space following the variety of the shapes one would like to reach. If the topology of the target shape is already known and if the available CAD parameter space is thought to be suitable, it should be considered as a control parameter during optimization. On the other hand, one might use a different parameter space, larger or smaller, during optimization having in mind that the final shape should be expressed in a usable CAD format. For some applications it is important to allow for variable topology; then shape parameters can be, for instance, a relaxed characteristic function (level set and immersed boundary approaches belong to this class). The different parameter spaces should be seen as complementary for primary and final stages of optimization. Indeed, the main advantage of a level set over a CAD-based parameter space is in primary optimization where the topology of the target shape is unknown and any a priori choice is hazardous.

An important issue in minimization is sensitivity evaluation. Gradients are useful in multi-criteria optimization to discriminate between Pareto equilibrium even when using gradient-free minimization algorithms. Sensitivities also permit us to introduce robustness issues during optimization. Robustness is also central in validation and verification as no simulation or design can be reliable if it is too sensitive to small perturbations in the data.

For sensitivity evaluation when the parameter space is large the most efficient approach is to use an adjoint variable with the difficulty that it requires the development of specific software. Automatic differentiation brings some simplification, but does not avoid the main difficulty of intermediate state storage, even though check-pointing techniques bring some relief. The use of commercial software without the source code is also a limitation for automatic differentiation and differentiable optimization in general. Incomplete sensitivity formulation and reduced order modelling are therefore preferred when possible to reduce this computational complexity and also because these often bring some extra understanding of the physics of the problem. These techniques are also useful for minimization with surrogate models as mentioned above.

Another important issue is that the results may depend on the evolution of the modelling. It is important to be able to provide information in an incremental way, following the evolution of the software. This means that we need the sensitivity of the result with respect to the different independent variables for the discrete form of the model and also that we need to be able to do that without re-deriving the model from scratch. But again use of commercial software puts serious limitations on what can be efficiently done and increases the need for adaptive reduced order modelling.

Hence, any design should be seen as constrained optimization. Adding robustness issues implies most likely a change in the solution procedures. From a practical point of view, it is clear that adding such requirements as those mentioned above will have a prohibitive cost, especially for multi-criteria and multi-physics designs. But answers are needed and even an incomplete study, even a posteriori, will require at least the sensitivity of the solution to perturbation of independent variables.

As implied by the title, this book deals with shape optimization problems for fluids with an emphasis on industrial application; it deals with the basic shape optimization problems for the aerodynamics of airplanes and some fluid-structure design problems. All these are of great practical importance in computational fluid dynamics (CFD), not only for airplanes but also for cars, turbines, and many other industrial applications. A new domain of application is also covered: shape optimization for microfluidic devices.

Let us also warn the reader that the book is not a synthesis but rather a collection of case studies; it builds on known material but it does not present this material in a synthetic form, for several reasons, like clarity, page numbers, and time. Furthermore a survey would be a formidable task, so huge is the literature.

So the book begins with a chapter on optimal shape design by local shape variations for simple linear problems, discretized by the finite element method. The goal is to provide tools to do the same with the complex partial differential equations of CFD. A general presentation of optimal shape design problems and of their solution by gradient algorithms is given. In particular, the existence of solutions, sensitivity analysis at the continuous and discrete levels are discussed, and the implementation problems for each case are pointed out. This chapter is therefore an introduction to the rest of the book. It summarizes the current knowhow for OSD, except topological optimization, as well as global optimization methods such as evolutionary algorithms.

In Chapter 3 the equations of fluid dynamics are recalled, together with the $k - \varepsilon$ turbulence model, which is used later on for high Reynolds number flows when the topology of the answer is not known. The fundamental equations of fluid dynamics are recalled; this is because applied OSD for fluids requires a good understanding of the state equation: Euler and Navier-Stokes equations in our case, with and without turbulence models together with the inviscid and/or incompressible limits. We recall wall functions also later used for OSD as low complexity models. By wall function we understand domain decomposition with a reduced dimension model near the wall. In other words, there is no universal wall function and when using a wall function, it needs to be compatible with the model used far from the wall. Large eddy simulation is giving a new life to the wall functions especially to simulate high-Reynolds external flows.

Chapter 4 deals with the numerical methods that will be used for the flow solvers. As in most commercial and industrial packages, unstructured meshes with automatic mesh generation and adaptation are used together with finite volume or finite element discretization for these complex geometries. The iterative solvers and the flux functions for upwinding are also presented here.

Then in Chapter 5 sensitivity analysis and automatic differentiation (AD) are presented: first the theory, then an object oriented library for automatic differentiation (AD) by operator overloading, and finally our experience with AD systems using code generation operating in both direct and reverse modes. We describe the different possibilities and through simple programs give a comprehensive survey of direct AD by operator overloading and for the reverse mode, the adjoint code method.

Chapter 6 presents parameterization and geometrical issues. This is also one of the key points for an efficient OSD platform. We describe different strategies for shape deformation within and without (level set and CAD-Free) computer aided design data structures during optimization. Mesh deformation and remeshing are discussed there. We discuss the pros and the cons of injection/suction boundary conditions equivalent to moving geometries when the motion is small. Some strategies to couple mesh adaptation and the shape optimization loop are presented. The aim is to obtain a multi-grid effect and improve convergence.

Chapter 7 gives a survey of optimization algorithms seen as discrete forms of dynamical systems. The presentation is not intended to be exhaustive but rather reflects our practical experience. Local and global optimizations are discussed. A unified formulation is proposed for both deterministic and stochastic optimizations. This formulation is suitable for multi-physic problems where a coupling between different models is necessary.

One important topic discussed in Chapter 8 is incomplete sensitivity. By incomplete sensitivity we mean that during sensitivity evaluation only the deformation of the geometry is accounted for and the change of the state variable due to the change of geometry is ignored. We give the class of functionals for which this assumption can be made. Incomplete sensitivity calculations are illustrated on several model problems. This gives the opportunity of introducing low-complexity models for sensitivity analysis. We show by experience that the accuracy is sufficient for quasi-Newton algorithms and also that the complexity of the method is drastically reduced making possible real time sensitivity analysis later used for unsteady applications.

In Chapter 9 we put forward a general argument to support the use of approximate gradients within optimization loops integrated with mesh refinements, although this does not justify all the procedures that are presented in Chapter 8. We also prove that smoothers are essential. This part was done in collaboration with E. Polak and N. Dicesare.

Then follows the presentation of some applications for stationary flows in Chapter 10 and unsteady problems in chapter 11. We gather in Chapter 10 examples of shape optimization in two and three space dimensions using the tools presented above for both inviscid and viscous turbulent cases. Chapter 11 presents applications of our shape optimization algorithms to cases where the flow is unsteady for rigid and elastic bodies and shows that control and OSD problems are particular cases of a general approach. Closed loop control algorithms are presented together with an analogy with dynamical systems.

Chapter 12 gathers the application of the ingredients above to the design of microfluidic devices. This is a new growing field. Most of what was made for aeronautics can be applied to these fluids at nearly zero Reynolds and Mach numbers.

The book closes with Chapter 13 on topological shape optimization described in simple situations.

The selection of this material corresponds to what we think to be a good compromise between complexity and accuracy for the numerical simulation of nonlinear industrial problems, keeping in mind practical aspects at each level of the development, illustrating our proposal, with many practical examples which we have gathered during several industrial cooperations. In particular, the concepts are explained more intuitively than with complete mathematical rigor. Thus this book should be important for whoever wishes to solve a practical OSD problem. In addition to the classical mathematical approach, the application of some modern techniques such as automatic differentiation and unstructured mesh adaptation to OSD are presented, and multi-model configurations and some time-dependent shape optimization problems are discussed.

The book has been influenced by the reactions of students who have been taught this material at the Masters level at the Universities of Paris and Montpellier. We think that what follows will be particularly useful for engineers interested in the implementation and solution of optimization problems using commercial packages, or in-house solvers, graduate and PhD students in applied mathematics, aerospace, or mechanical engineering needing, during their training and research, to understand and solve design problems, and research scientists in applied mathematics, fluid dynamics, and CFD looking for new exciting research and development areas involving realistic applications.

2
OPTIMAL SHAPE DESIGN

2.1 Introduction

In mathematical terms, an optimal shape design (OSD) requires the optimization of one or several criteria $\{E_i(x)\}_1^I$ which depend on design parameters $x \in X$ which define the shape of the system within an admissible set of values X.

Multi-criteria optimization is a difficult field in itself of which we shall retain only the min-max idea:

$$\min_{x \in X}\{J(x) \; : \; E_i(x) \leq J(x), \quad i = 1, ..., I\}.$$

By definition a Pareto optimum $x \in X$ is such that there is no $y \in X$ such that $E_i(y) < E_i(x)$ for all $i \in I$. For convex functionals E_i and convex X, it can be shown that a Pareto point also solves

$$\min_{x \in X} \sum_{i=1}^{I} \alpha_i E_i(x).$$

For some suitable values of $\alpha_i \in (0,1)$, both problems are related and solve in some intuitive sense the multi-criteria problem. Differentiable optimization and control theory is easily applied to these derived single criteria problems.

Optimal control for distributed systems [35] is a branch of optimization for problems which involve a parameter or control variable u, a state variable y, and a partial differential equation (PDE) A (with boundary conditions b), to define y in a domain Ω:

$$\min_{u,y}\{J(u,y) \; : \; A(x,y,u) = 0 \; \forall x \in \Omega, \; b(x,y,u) = 0 \; \forall x \in \partial\Omega\}.$$

For example,

$$\min_{u,y}\left\{\int_\Omega (y-1)^2 \; : \; -\Delta y = 0 \; \forall x \in \Omega, \; y|_{\partial\Omega} = u\right\}, \tag{2.1}$$

attempts to find a boundary condition u for which y would be as close to the value 1 as possible.

For (2.1) there is a trivial unique solution $u = 1$ because $y = 1$ is a solution to the Laplace equation.

Optimal shape design is a special case of control theory where the control is the boundary $\partial\Omega$ itself. For example, if D is given, consider

$$\min_{\{\partial\Omega, D\subset\Omega\}} \left\{ \int_D (y-1)^2 \; : \; -\Delta y = g, \; y|_{\partial\Omega} = 0 \right\}. \tag{2.2}$$

When $g = -9.81$ this problem has a physical meaning [51]; it attempts to answer the following: is it possible to build a support for a membrane bent by its own weight which would bring its deflection as close to 1 as possible in a region of space D?

The intuitive answer is that $y = 1$ in D is impossible unless $g = 0$ and indeed it is incompatible with the PDE as $-\Delta 1 = 0 \neq g$. If $g = 0$ then $y = 0$ everywhere and so the objective function is always equal to the area of D for any Ω containing D : every shape is a solution.

Thus uniqueness of the solution can be a problem even in simple cases.

In smooth situations, the PDE can be viewed as an implicit map $u \to \Omega(u) \to y(u)$ where $u \to \Omega(u)$ is the parameterization of the domain by a (control) parameter u and the problem is to minimize the function $J(u, y(u))$. If it is continuously differentiable in u, then the algorithms of differentiable optimization can be used (see [49] for instance) and so it remains only to explain how to compute J'_u.

Analytic computation of derivatives for OSD problems is possible both for continuous and discretized problems. It may be tedious but it is usually possible. When it is difficult one may turn to automatic differentiation (AD), but then other difficulties pop up and so it is a good idea to understand the theory even when using AD.

Therefore we begin this chapter by giving simple examples of OSD problems. Then we recall some theorems on the existence of solutions and give, for simple cases, a method to derive optimality conditions. Finally, we show the same on OSD problems discretized by the finite element method of degree one on triangulations. More details can be found in [11, 25–27, 33, 36, 45, 42, 44, 47].

2.2 Examples

2.2.1 Minimum weight of structures

In 2D linear elasticity, for a structure clamped on $\Gamma = \partial\Omega$, and subject to volume forces F, the horizontal displacement $u = (u_1, u_2) \in V$ is found by solving:

$$\int_\Omega \tau_{ij}(u)\epsilon_{ij}(v) = \int_\Omega F.v \quad \forall v \in V_0 = \{u \in H^1(\Omega)^2 \; : \; u|_\Gamma = 0\}$$

where $\epsilon_{ij}(u) = \frac{1}{2}(\partial_i u_j + \partial_j u_i)$, and

$$\begin{pmatrix} \tau_{11} \\ \tau_{22} \\ \tau_{12} \end{pmatrix} = \begin{pmatrix} 2\mu+\lambda & \lambda & 0 \\ \lambda & 2\mu+\lambda & 0 \\ 0 & 0 & 2\mu \end{pmatrix} \begin{pmatrix} \epsilon_{11} \\ \epsilon_{22} \\ \epsilon_{12} \end{pmatrix}$$

and where λ, μ are the Lamé coefficients.

Many important problems of design arise when one wants to find the structure with minimum weight yet satisfying some inequality constraints for the stress such as in the design of light weight beams for strengthening airplane floors, or for crankshaft weight optimization.

For all these problems the criterion for optimization is the weight

$$J(\Omega) = \int_\Omega \rho(x)dx,$$

where $\rho(x)$ is the density of the material at $x \in \Omega$. But there are constraints of the type

$$\tau(x) \cdot d(x) < \tau_{d\max},$$

at some points x and for some directions $d(x)$.

Indeed, a wing, for instance, needs to have a different response to stress spanwise and chord-wise. Moreover, due to coupling between physical phenomena, the surface stresses come in part from fluid forces acting on the wing. This implies many additional constraints on the aerodynamic (drag, lift, moment) and structural (Lamé coefficients) characteristics of the wing. Therefore, the Lamé equations of the structure must be coupled with the equations for the fluid (fluid structure interactions). This is why most optimization problems nowadays require the solution of several state equations, fluid and structure in this example.

2.2.2 Wing drag optimization

An important industrial problem is the optimization of the shape of a wing to reduce the drag. The drag is the reaction of the flow on the wing; its component in the direction of flight is the drag proper and the rest is the lift. A few percent of drag optimization means a great saving on commercial airplanes.

For viscous drag the Navier-Stokes equations must be used. For wave drag the Euler system is sufficient.

For a wing S moving at constant speed u_∞ the force acting on the wing is

$$F = \int_S \left[\mu(\nabla u + \nabla u^T) - \frac{2\mu}{3}\nabla \cdot u\right] n - \int_S pn,$$

where n is the normal to S pointing outside the domain occupied by the fluid.

The first integral is a viscous force, the so-called viscous drag/lift, and the second is called the wave drag/lift. In a frame attached to the wing, and with uniform flow at infinity, the drag is the component of F parallel to the velocity at infinity (i.e. $F.u_\infty$). The viscosity of the fluid is μ and p is its pressure.

The Navier-Stokes equations govern u the fluid velocity, θ the temperature, ρ the density and E the energy:

$$\partial_t \rho + \nabla.(\rho u) = 0,$$

$$\partial_t(\rho u) + \nabla.(\rho u \otimes u) + \nabla p - \mu \Delta u - \frac{1}{3}\mu \nabla(\nabla.u) = 0,$$

$$\partial_t[\rho E] + \nabla \cdot [u\rho E] + \nabla \cdot (pu)$$
$$= \nabla \cdot \{\kappa \nabla \theta + [\mu(\nabla u + \nabla u^T) - \frac{2}{3}\mu)\mathbf{I}\nabla \cdot u]u\},$$

where $E = \dfrac{u^2}{2} + \theta$ and $p = (\gamma - 1)\rho\theta$.

The Euler equations are obtained by letting $\kappa = \mu = 0$ in the Navier-Stokes equations.

The problem is to minimize

$$J(S) = F.u_\infty,$$

with respect to the shape of S.

There are several *constraints*: a geometrical constraint such as the volume being greater than a given value, else the solution will be a point; and, an aerodynamic constraint: the lift must be greater than a given value or the wing will not fly.

The problem is difficult because it involves the Navier-Stokes equations at high Reynolds number. It can be simplified by considering only the wave drag, i.e. the pressure term only in the definition of F [32]. Then the viscous terms can be dropped in the Navier-Stokes equations ($\mu = \kappa = 0$); Euler's equations remain.

However, there may be side effects to such simplifications. In transonic regimes, for instance, the "shock" position for a Navier-Stokes flow is upstream compared to an inviscid (Euler) simulation at the same Mach number. Figures 2.1 and 2.3 display the results of two optimizations using Euler equations and a Navier-Stokes equations with $k - \varepsilon$ turbulence modeling for a NACA 0012 at Mach number of 0.75 and 2° of incidence. One aims at reducing the drag at given lift and volume.

Simplifying the state equation Assuming irrotational flow an even greater simplification replaces the Euler equations by the compressible potential equation:

$$u = \nabla\varphi, \quad \rho = (1 - |\nabla\varphi|^2)^{1/(\gamma-1)}, \quad p = \rho^\gamma, \quad \nabla.(\rho u) = 0,$$

or even, if at low Mach number, by the incompressible potential flow equation:

$$u = \nabla\varphi, \quad -\Delta\varphi = 0. \tag{2.3}$$

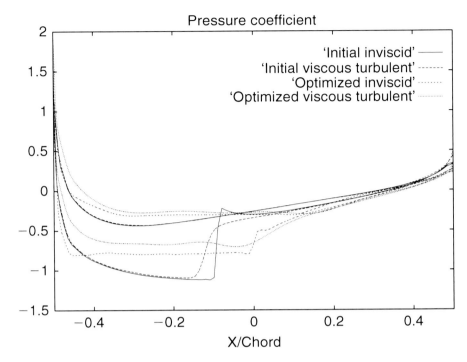

FIG. 2.1. Transonic drag reduction. Pressure coefficient.

The pressure is given by the Bernoulli law $p = p_{\text{ref}} - \frac{1}{2}u^2$ and so only an optimization of the lift would be :

$$\min_{S \in G} - \int_S f\left(\frac{u^2}{2}\right) n \cdot u_\infty^\perp \quad \text{subject to (2.3) and}$$

$$\frac{\partial \phi}{\partial n}|_{\Gamma - S} = u_\infty \cdot n, \quad \frac{\partial \phi}{\partial n}|_S = 0,$$

for some admissible set of shapes G and some local criteria f.

Multi-point optimization Engineering constraints on admissible shapes are numerous: minimal thickness, given length, maximum admissible radius of curvature, minimal angle at the trailing edge.

Another problem arises due to instability of optimal shapes with respect to data. It has been seen that the leading edge at the optimum is a wedge. Thus if the incidence angle for u_∞ is changed the solution becomes bad. A multi-point functional must be used in the optimization:

$$\min_S J(S) = \sum \alpha_i u_\infty^i \cdot F^i \quad \text{or} \quad \min\{J \ : \ u_\infty^i \cdot F^i \leq J \ \forall i\},$$

at given lifts where the F^i are computed from the Navier-Stokes equations with boundary conditions $u = u_\infty^i$, $u|_S = 0$.

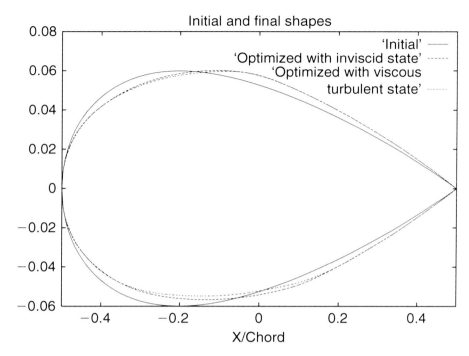

FIG. 2.2. Transonic drag reduction. Initial and final shapes for inviscid and viscous optimizations. The differences between the two shapes increase with the deviation between the shock positions.

2.2.3 *Synthetic jets and riblets*

The solution to a time dependent optimization problem is time dependent. But for wings this would give a deformable shape, with motion at the time scale of the turbulence in (2.3). As this is computationally unreachable, suboptimal solutions may be sought.

One direction is to replace moving surfaces by mean surfaces which can breathe. For instance, consider a surface with tiny holes each connected to a rubber reservoir activated by an electronic device capable therefore of blowing and sucking air so long as the net flow is zero over a period. The reservoir may be ignored and the mean surface may be considered with transpiration conditions [22, 38, 41].

In the class of time-independent shapes with time-dependent flows it is not even clear that the solution is smooth. In [22], the authors showed that riblets, little groves in the direction of the flow, actually reduce the drag by a few percent. The simulation was done with a large eddy simulation (LES) model for the turbulence and at the time of writing this book shape optimization with LES is beyond our computational power. But this is certainly an important research area for the future.

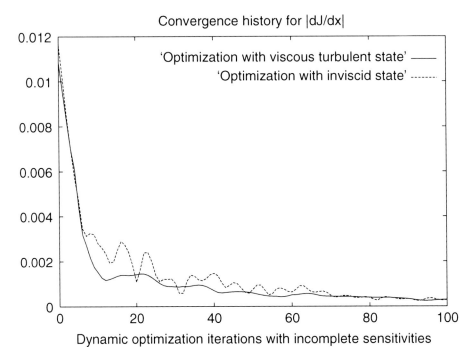

FIG. 2.3. Transonic drag reduction. Convergence histories for the gradients using inviscid and viscous flows. The convergence seems to be more regular the viscous flows: with a robust solver for the turbulence model, optimization is actually easier than with the Euler equations. Of course, the CPU time is larger because the viscous case requires a finer mesh.

2.2.4 Stealth wings

Maxwell equations The optimization of the far-field energy of a radar wave reflected by an airplane in flight requires the solution of Maxwell's equations for the electric field E and the magnetic field H:

$$\epsilon \partial_t E + \nabla \times H = 0, \quad \nabla . E = 0, \quad \mu \partial_t H - \nabla \times E = 0, \quad \nabla . H = 0.$$

The electric and magnetic coefficients ϵ, μ are constant in air but not so in an absorbing medium. One variable, H for instance, can be eliminated by differentiating the first equation with respect to t:

$$\epsilon \partial_{tt} E + \nabla \times (\frac{1}{\mu} \nabla \times E) = 0.$$

It is easy to see that $\nabla . E = 0$ if it is zero at the initial time.

Helmholtz equation Now if the geometry is cylindrical with axis z and if $E = (0, 0, E_z)^T$ then the equation becomes a scalar wave equation for E_z.

Furthermore, if the boundary conditions are periodic in time at infinity, $E_z = \mathcal{R}_e v_\infty e^{i\omega t}$ and compatible with the initial conditions then the solution has the form $E_z = \mathcal{R}_e v(x) e^{i\omega t}$ where v, the amplitude of the wave E_z of frequency ω, is the solution of:

$$\nabla \cdot (\frac{1}{\mu} \nabla v) + \omega^2 \epsilon v = 0. \tag{2.4}$$

Notice the incorrect sign for the ellipticity in the Helmholtz equation (2.4). This equation also arises in acoustics. In vacuum $\mu\epsilon = c^2$, c the speed of light, so for numerical purposes it is a good idea to re-scale the equation. The critical parameter is then the number of waves on the object, i.e. $\omega c/(2\pi L)$ where L is the size of the object.

Boundary conditions The reflected signal on solid boundaries S satisfies

$$v = 0 \text{ or } \partial_n v = 0 \text{ on } S,$$

depending on the type of waves (transverse magnetic polarization requires Dirichlet conditions).

When there is no scattering object this Helmholtz equation has a simple sinusoidal set of solutions which we call v_∞:

$$v_\infty(x) = \alpha \sin(k \cdot x) + \beta \cos(k \cdot x),$$

where k is any vector of modulus $|k| = \omega c$. Radar waves are more complex, but by Fourier decomposition they can be viewed as a linear combination of such simple unidirectional waves.

Now if such a wave is sent onto an object, it is reflected by it and the signal at infinity is the sum of the original and the reflected waves. So it is better to set an equation for the reflected wave only $u = v - v_\infty$.

A good boundary condition for u is difficult to set; one possibility is

$$\partial_n u + iau = 0.$$

Indeed, when $u = e^{id \cdot x}$, $\partial_n u + iau = i(d \cdot n + a)u$, so that this boundary condition is transparent to waves of direction d when $a = -d \cdot n$. If we want this boundary condition to let all *outgoing* waves pass the boundary, we will set $a = -ik \cdot n$.

To summarize, we set, for u, the system in the complex plane:

$$\nabla \cdot \left(\frac{1}{\mu} \nabla u\right) + \omega^2 u = 0, \text{ in } \Omega,$$
$$\partial_n u - ik \cdot nu = 0 \text{ on } \Gamma_\infty,$$
$$u = g \equiv -e^{ik \cdot x} \text{ on } S,$$

where $\partial\Omega = S \cup \Gamma_\infty$. It can be shown that the solution exists and is unique. Notice that the variables have been rescaled, ω is ωc, μ is μ/μ_{vacuum}.

Usually the criterion for optimization is a minimum amplitude for the reflected signal in a region of space D at infinity (hence D is an angular sector). For instance, one can consider

$$\min_{S \in \mathcal{O}} \int_{\Gamma_\infty \cap D} |\nabla u|^2 \quad \text{subject to} \tag{2.5}$$

$$\omega^2 u + \nabla \cdot (\frac{1}{\mu} \nabla u) = 0, \quad u|_S = g, \quad (-ik \cdot nu + \partial_n u)|_{\Gamma_\infty} = 0.$$

In practice μ is different from 1 only in a region very close to S so as to model the absorbing paint that most stealth airplanes have.

But constraints are aerodynamic as well, lift above a given lower limit for instance, and thus require the solution of the fluid part as well.

The design variables are the shape of the wing, the thickness of the paint, the material characteristics (ϵ, μ) of the paint.

Here, again, the theoretical complexity of the problem can be appreciated from the following question: would riblets on the wing, of the size of the radar wave, improve the design?

Homogenization can answer the question [1, 6, 7]; it shows that an oscillatory design is indeed better. Furthermore, periodic surface irregularities are equivalent, in the far field, to new effective boundary conditions:

$$u = 0 \quad \text{on an oscillatory } S^\epsilon \text{ can be replaced by}$$

$$au + \partial_n u = 0 \quad \text{on a mean } S,$$

for some suitable a [2].

If that is so then the optimization can be done with respect to a only. But optimization with respect to the parameters of the PDE is known to generate oscillations [55]; this topic is known as topological optimization (see below).

Optimization with respect to μ also gives rise to complex composite structure design problems.

So an aerodynamic constraint on the lift has been added. The flow is assumed inviscid and irrotational and computed by a stream function ψ:

$$u = \nabla \times \psi, \quad p = p_{\text{ref}} - \frac{u^2}{2}, \quad \Delta \psi = 0 \text{ in } \Omega,$$

$$\psi|_S = \beta \quad \psi|_{\Gamma_\infty} = \begin{pmatrix} \cos \theta \\ \sin \theta \end{pmatrix} \times n,$$

where u, p are the velocity and pressure in the flow, S is the wing profile, θ its angle of incidence, and n its normal. The constant β is adjusted so that the pressure is continuous at the trailing edge [48].

The lift being proportional to β we impose the constraint $\beta \geq \beta_0$ the lift of the NACA0012 airfoil. The result after optimization is shown in Fig. 2.4. Without constraint the solution is very unaerodynamic.

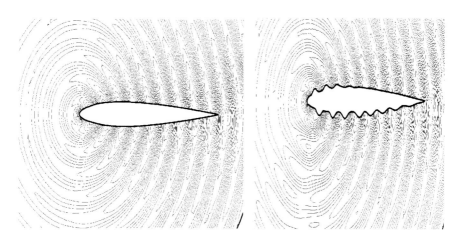

FIG. 2.4. Stealth wing. Optimization without and with aerodynamic constraint. (Courtesy of A. Baron, with permission.)

2.2.5 Optimal breakwater

Here the problem is to build a calm harbor by designing an optimal breakwater [12, 30, 31]. As a first approximation, the amplitude of sea waves satisfies the Helmholtz equation

$$\nabla \cdot (\mu \nabla u) + \epsilon u = 0, \tag{2.6}$$

where μ is a function of the water depth and ϵ is function of the wave speed.

With approximate reflection and damping whenever the waves die out on the coast or collide on a breakwater S which is surrounded by rocks, we have

$$\partial_n u + au = 0 \quad \text{on} \quad S. \tag{2.7}$$

for some appropriate a.

At infinity a non-reflecting boundary condition must be used, for instance

$$\partial_n(u - u_\infty) + ia(u - u_\infty) = 0. \tag{2.8}$$

The problem is to find the best S with given length and curvature constraints so that the waves have minimum amplitude in a given harbor D:

$$\min_{\partial\Omega} \int_D |u|^2 : \text{ subject to } (2.6), (2.7), (2.8). \tag{2.9}$$

To illustrate the chapter we show the results of the theory on an example.

We seek the best breakwater to protect a harbor of size L from uniformly sinusoidal waves at infinity of wave length $\lambda \times L$ and direction α. The amplitude of the wave is given by the Helmholtz equation (2.6). The real part of the amplitude is shown in Fig. 2.5 before optimization (a straight dyke) and after optimization. Constraints on the length of the dyke and on its monotonicity have been imposed.

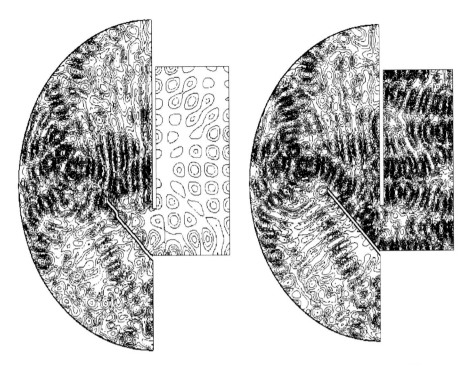

FIG. 2.5. Breakwater optimization. The aim is to get a calm harbor. (Courtesy of A. Baron, with permission.)

2.2.6 *Two academic test cases: nozzle optimization*

For clarity we will explain the theory on a simple optimization problem for incompressible irrotational inviscid flows. The problem is to design a nozzle so as to reach a desired state of velocity u_d in a region of space D. Incompressible irrotational flow can be modeled with a potential function φ in which case the problem is

$$\min_{\partial\Omega} \int_D |\nabla\varphi - u_d|^2 \; : \; -\Delta\varphi = 0 \text{ in } \Omega, \; \partial_n\varphi|_{\partial\Omega} = g, \qquad (2.10)$$

or with a stream function in 2D, in which case we must solve

$$\min_{\partial\Omega} \int_D |\nabla\psi - v_d|^2 \; : \; -\Delta\psi = 0 \text{ in } \Omega, \; \psi|_{\partial\Omega} = \psi_\Gamma. \qquad (2.11)$$

In both problems one seeks a shape which produces the closest velocity to u_d in the region D of Ω. In the second formulation the velocity of the flow is given by:

$$(\partial_2\psi, -\partial_1\psi)^T, \quad \text{so} \quad v_d = (u_{d2}, -u_{d1})^T.$$

The difference between the two problems is in the boundary condition, Neumann in the first case and Dirichlet in the second.

Application to wind tunnel or nozzle design for potential flow is obvious but it is academic because these are usually used with compressible flows.

2.3 Existence of solutions

2.3.1 Topological optimization

In a seminal paper, Tartar [55] showed that the solution u^n of a PDE like

$$\mathcal{L}(a^n)u := u - \nabla \cdot (a^n \nabla u) = f, \quad u \in H_0^1(\Omega),$$

may not converge to the solution of $\mathcal{L}(a^*)u = f$ when $\lim_{n\to\infty} a^n = a^*$. In many cases the limit u^* would be solution of

$$u - \nabla \cdot (A \nabla u) = f, \quad u \in H_0^1(\Omega),$$

where A is a matrix. Homogenization theory [14] gives tools to find A.

The same phenomena can be observed with domains Ω^n. Indeed, let $\hat{\Omega}$ be a reference domain and consider the set of domains $T(\hat{\Omega})$ where the mapping $T : \mathcal{R}^d \to \mathcal{R}^d$ has bounded first derivatives. Further, assume that for some domain D: $T(x) = x \ \forall x \in D$. Define

$$\mathcal{O} = \{T(\hat{\Omega}) \ : \ T\mathcal{R}^d \to \mathcal{R}^d, \ T \in W_\infty^1(\mathcal{R})^d, \ T(D) = D\}.$$

Recall that

$$\int_\Omega \nabla y \nabla z \, dx = \int_{\hat{\Omega}} \nabla y \cdot T'^{-1} T'^{-T} \nabla z \det T' \, d\hat{x}.$$

Therefore, problem (2.2) for instance, is also

$$\min_{T \in \mathcal{O}} \int_D (y-1)^2 d\hat{x} \ : \tag{2.12}$$

$$\int_{\hat{\Omega}} [\nabla y \cdot T'^{-1} T'^{-T} \nabla z - gz] \det T' d\hat{x} = 0 \ \forall z \in H_0^1(\hat{\Omega}). \tag{2.13}$$

When (2.12) is solved by an optimization algorithm a minimizing sequence T^n is generated, but the limit of T^n may not be an element of $\mathcal{O}$. This happens, for instance, if the boundary of Ω^n oscillates with finer and finer oscillations of finite amplitude or if Ω^n has more and more holes.

Topological optimization [3–5,13] studies the equations of elasticity (and others) in media made of two materials or made of material with holes. It shows that a material with infinitely many small holes may have variable Lamé coefficients in the limit for instance.

Consequently if oscillations occur in the numerical solution of (2.12) it may be because it has no solution or because the numerical method does not converge; then one should include more constraints on the admissible domains to avoid oscillations and holes (bounded radius of curvature for instance) or solve the relaxed problem $\nabla \cdot (A\nabla z) = 0$. In this book we will not study relaxed problems but we will study the constraints, and/or Tikhonov regularizations, which can be added to produce a well-posed problem.

2.3.2 Sufficient conditions for existence

For simplicity we translate the non-homogeneous boundary conditions of the academic example (2.11) above into a right-hand side in the PDE ($f = \Delta\psi_\Gamma$). So denote by $\psi(\Omega)$ the solution of

$$-\Delta\psi = f \text{ in } \Omega, \quad \psi|_{\partial\Omega} = 0.$$

Assume that $u_d \in L^2(\Omega)$, $f \in H^{-1}(\Omega)$. Let $O \supset D$ be two given closed bounded sets in $R^d, d = 2$ or 3 and consider

$$\min_{\Omega \in \mathcal{O}} J(\Omega) = \int_D |\nabla\psi(\Omega) - u_d|^2,$$

with

$$\mathcal{O} = \{\Omega \subset R^d \ : \ O \supset \Omega \supset D, \ |\Omega| = 1\},$$

where $|\Omega|$ denotes the area in 2D and the volume in 3D.

In [21], it is shown that there exists a solution provided that the class $\mathcal{O}$ is restricted to Ω's which are locally on one side of their boundaries, and satisfy the *Cone Property*, i.e. that $D_\epsilon(x,d)$, the intersection with the sphere of radius ϵ and center x of the cone of vertex x direction d and angle ϵ is such that:

Cone property: *There exists ϵ such that for every $x \in \partial\Omega$ there exists d such that $\Omega \supset D_\epsilon(x,d)$.*

These two conditions imply that the boundary cannot oscillate too much.

In 2D an important result of existence has been obtained by Sverak [53] under very mild constraints.

Theorem 2.1 *Let $\mathcal{O} = \mathcal{O}_N$ be the set of open sets Ω containing D and included in O, possibly with a constraint on the area such as $|\Omega| \geq 1$, and which have less than N connected components. The problem*

$$\min_{\mathcal{O}_N} J(\Omega) = \int_D |\nabla\psi(\Omega) - v_d|^2 \ : \ -\Delta\psi(\Omega) = f \text{ in } \Omega, \ \psi(\Omega)|_{\partial\Omega} = 0$$

has a solution.

In other words, two things can happen to minimizing sequences. Either accumulation points are solutions, or the number of holes in the domain tends to infinity (and their size to zero).

This result is false in 3D as it is possible to make shapes with spikes such that a 2D cut will look like a surface with holes and yet the 3D surface remains singly connected. An extension of the same idea can be found in [18] with the flat cone hypothesis: *If the boundary of the domain has the flat cone insertion property (each boundary point is the vertex of a fixed size 2D truncated cone which fits inside the domain) then the problem has at least one solution.*

2.4 Solution by optimization methods

Throughout this section we will assume that the cost functions of the optimization problems are continuously differentiable: let V be a Banach space and $v \in V \to J(v) \in \mathcal{R}$. Then $J'_v(v)$ is a linear operator from V to $\mathcal{R}$ such that

$$J(v + \delta v) = J(v) + J'_v(v)\delta v + o(\|\delta v\|).$$

2.4.1 Gradient methods

At the basis of gradient methods is the Taylor expansion for

$$J(v + \lambda w) = J(v) + \lambda \langle \mathrm{Grad}_v J, w \rangle + o(\lambda \|w\|), \quad \forall v, w \in V, \quad \forall \lambda \in \mathcal{R}, \quad (2.14)$$

where V is now a Hilbert space with scalar product $\langle \cdot, \cdot \rangle$ and $\mathrm{Grad}_v J$ is the element of V given by the Ritz theorem and defined by

$$\langle \mathrm{Grad}_v J, w \rangle = J'_v w, \quad \forall w \in V.$$

By taking $w = -\rho \mathrm{Grad}_v J(v)$ in (2.14), with $0 < \rho << 1$ we find:

$$J(v + w) - J(v) = -\rho \|\mathrm{Grad}_v J(v)\|^2 + o(\rho \|\mathrm{Grad}_v J(v)\|).$$

Hence if ρ is small enough the first term on the right-hand side will dominate the remainder and the sum will be negative:

$$\rho \|\mathrm{Grad}_v J(v)\|^2 > o(\rho \|\mathrm{Grad}_v J(v)\|) \quad \Rightarrow \quad J(v + w) < J(v).$$

Thus the sequence defined by :

$$v^{n+1} = v^n - \rho \mathrm{Grad}_v J(v), \quad n = 0, 1, 2, \ldots \quad (2.15)$$

makes $J(v^n)$ monotone decreasing. We have the following result:

Theorem 2.2 *If J is continuously differentiable, bounded from below, and $+\infty$ at infinity, then all accumulation points v^* of v^n, generated by (2.15) satisfy*

$$\mathrm{Grad}_v J(v^*) = 0.$$

This is the so-called *optimality condition* of order 1 of the problem. If J is convex then it implies that v^* is a minimum; if J is strictly convex the minimum is unique.

By taking the best ρ in the *direction of descent* $w^n = -\mathrm{Grad}_v J(v^n)$,

$$\rho^n = \mathrm{argmin}_\rho J(v^n + \rho w^n), \quad \text{meaning that}$$

$$J(v^n + \rho^n w^n) = \min_\rho J(v^n + \rho w^n).$$

We obtain the so-called *method of steepest descent with optimal step size*.

We have to remark, however, that minimizing a one parameter function is not all that simple. The exact minimum cannot be found in general, except for polynomial functions J. So in the general case, several evaluations of J are required for the quest of an approximate minimum only.

A closer look at the convergence proof of the method [49] shows that it is enough to find ρ^n with the gradient method with Armijo rule:

- **Choose** $v^0, 0 < \alpha < \beta < 1$;
- **Loop on n**
  * Compute $w = -\mathrm{Grad}_v J(v^n)$,
  * Find ρ such that
  $$-\rho\beta\|w\|^2 < J(v^n + \rho w) - J(v^n) < -\rho\alpha\|w\|^2,$$
  * Set $v^{n+1} = v^n + \rho w$,
- **end loop**

An approximate Armijo rule takes only one line of slope $\alpha\|w\|^2$ and first finds the largest ρ of the form $\rho = \rho_0 2^{\pm k}$ which gives a decrement for J below the line for ρ and above the line for 2ρ:

Choose $\rho_0 > 0$, $\alpha \in (0,1)$ and find $\rho = \rho_0 2^k$ where k is the smallest signed integer (k can be negative) such that

$$J(v^n + \rho w) - J(v^n) < -\rho\alpha\|w\|^2 \quad \text{and} \quad -2\rho\alpha\|w\|^2 \leq J(v^n + 2\rho w) - J(v^n).$$

A good choice is $\rho^0 = 1$, $\alpha = 1/2$. Another way is to compute ρ_1^n, ρ_2^n iteratively from an initial guess ρ and $\rho_1^0 = \rho, \rho_2^0 = 2\rho$ by $\rho_i^{n+1} = (\rho_1^n + \rho_2^n)/2$ or $\rho_i^{n+1} = \rho_i^n$ ($i = 1, 2$) such that ρ_1^{n+1} produces an increment below the line and ρ_2^{n+1} above it.

2.4.2 Newton methods

The method of Newton with optimal step size applied to the minimization of J is

$$w \text{ solution of } J''_{vv} w = -\mathrm{Grad}_v J(v^n),$$
$$v^{n+1} = v^n + \rho^n w^n,$$
$$\text{with } \rho^n = \arg\min_\rho J(v^n + \rho w).$$

Near to the solution it can be shown that $\rho^n \to 1$ so that it is also the root-finding Newton method applied to the optimality condition

$$\mathrm{Grad}_v J(v) = 0.$$

It is quadratically convergent but it is expensive and usually J'' is difficult to compute, so a quasi-Newton method, where an approximation of J'' is used, may be preferred. For instance, an approximation to the exact direction w can be found by:

Choose $0 < \epsilon \ll 1$, compute an approximate solution w of
$$\frac{1}{\epsilon}(\mathrm{Grad}_v J(v^n + \epsilon w) - \mathrm{Grad}_v J(v^n)) = -\mathrm{Grad}_v J(v^n).$$

2.4.3 Constraints

The simplest method (but not the most efficient) to deal with constraints is by penalty. Consider the following minimization problem in $x \in R^N$ under equality and inequality constraints on the control x and state u:

$$\min_x J(x, u(x)) : \quad \text{with } A(x, u(x)) = 0, \text{ and also subject to}$$
$$B(x, u(x)) \leq 0, \quad C(x, u(x)) = 0, \quad x_{\min} \leq x \leq x_{\max}. \tag{2.16}$$

Here A is the state equation and B and C are vector-valued constraints on x and u while the last inequalities are box constraints on the control only.

The problem can be approximated by penalty

$$\min_{x_{\min} \leq x \leq x_{\max}} \{E(x) = J(x, u(x)) + \beta |B^+|^2 + \gamma |C|^2 : A(x, u(x)) = 0\},$$

where β and γ are penalty parameters, which, it must be stressed, are usually difficult to choose in practice because the theory requires that they tend to infinity but the conditioning of the problem deteriorates when they are large.

Calculus of variation gives the change in the cost function due to a change $x \to x + \delta x$:

$$E(x + \delta x) - E(x) \simeq (J'_x + 2\beta {B^+}^T B'_x + 2\gamma C^t C'_x) \delta x$$
$$+ (J'_u + 2\beta {B^+}^T B'_u + 2\gamma C^t C'_u) \delta u,$$

where $\simeq$ stands for equality up to higher order terms. But,

$$A'_u \delta u + A'_x \delta x \simeq 0 \quad \Longrightarrow \quad \delta u \simeq - {A'_u}^{-1} A'_x \delta x.$$

So, we find that by defining p by

$$ {A'_x}^T p = \text{Grad}_u J + 2\beta \text{Grad}_u B B^+ + 2\gamma \text{Grad}_u C\ C.$$

We have

$$(J'_u + 2\beta {B^+}^T B'_u + 2\gamma C^t C'_u) \delta u = \delta u . {A'_x}^T p = A'_x \delta u . p = -p^T A'_x \delta x,$$
$$E'_x = J'_x + 2\beta {B^+}^T B'_x + 2\gamma C C'_x - p^T A'_x.$$

At each iteration of the gradient method, the new prediction is kept inside this box by projection:

$$x^{n+\frac{1}{2}} = x^n - \rho(\text{Grad}_x J + 2\beta \text{Grad}_x B B^+ + 2\gamma \text{Grad}_x C C - \text{Grad}_x A p),$$
$$x^{n+1} = \min(\max(x^{n+\frac{1}{2}}, x_{\min}), x_{\max}),$$

where the min and max and applied on each component of the vectors.

2.4.4 A constrained optimization algorithm

The following constraint optimization algorithm [29] has been fairly efficient on shape optimization problems [12,23,34]. It is a quasi-Newton method applied to the Lagrangian. Consider

$$\min_x \{J(x) : B(x) \leq 0\}.$$

It has for (Kuhn-Tucker) optimality conditions:

$$J'(x) + B'(x).\lambda = 0,$$
$$B(x).\lambda = 0 \quad B(x) \leq 0 \quad \lambda \geq 0.$$

Apply a Newton step to the equalities:

$$\begin{pmatrix} J'' + B''.\lambda & B'^T \\ B'.\lambda & B \end{pmatrix} \begin{pmatrix} x^{n+1} - x^n \\ \lambda^{n+1} - \lambda^n \end{pmatrix} = - \begin{pmatrix} J' + B'.\lambda^n \\ B.\lambda^n \end{pmatrix} \simeq \begin{pmatrix} -J' \\ -\rho(\lambda^n.e)e \end{pmatrix},$$

where $e = (1, 1, ..., 1)^T$ because $B_i(x^n) \leq 0$, $\lambda_i \geq 0$ implies $0 \geq \sum B'_i(x^n).\lambda_i$ which is crudely approximated by $-\rho \sum \lambda_j$ for some $\rho > 0$.

Notice that it is an interior algorithm: $(B(x^n) \leq 0$ for all $n)$.

2.5 Sensitivity analysis

In functional spaces, as in finite dimension, gradient and Newton methods require the gradient of the cost function J and for this we need to define an underlying Hilbert structure for the parameter space, the shapes. Several ways have been proposed:

Assume that all admissible shapes are obtained by mapping on a reference domain $\hat{\Omega}$: $\Omega = T(\hat{\Omega})$. Then the parameter is $T : \mathcal{R}^d \to \mathcal{R}^d$. A possible Hilbert space for T is the Sobolev space of order m and it seems that $m = 2$ is a good choice because it is close to $W^{1,\infty}$ (see [39]).

For practical purposes it is not so much the Hilbert structure of the space of shapes which is important, but the Hilbert structure for the tangent plane of the parameter space, meaning by this that the scalar product is needed only for small variations of $\partial\Omega$. So one works with local variations defined around a reference boundary Σ by

$$\Gamma(\alpha) = \{x + \alpha(x)n_\Sigma(x) \ : \ x \in \Sigma\},$$

where n_Σ is the outer normal to Σ at x and Ω is the domain which is on the left side of the oriented boundary $\Gamma(\alpha)$. Then the Hilbert structure is placed on α, for instance $\alpha \in H^m(\Sigma)$.

Similarly with the mappings to a reference domain it is possible to work with a local (tangent) variation $tV(x)$ and set

$$\Omega(tV) = \{x + tV(x) \ : \ x \in \Omega\} \quad t \text{ small and constant}, \qquad (2.17)$$

where $tV(x)$ is the infinitesimal domain displacement at x.

One can also define the shape as the zero of a *level set function* ϕ:

$$\Omega = \{x \; : \; \phi(x) \leq 0\}.$$

Then the unknown is ϕ for which there is a natural Hilbert structure, for instance in 2D, the Sobolev space H^2 because the continuity of ϕ is needed (see Fig. 2.6).

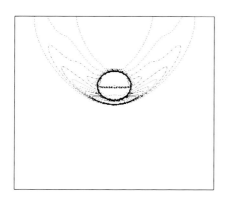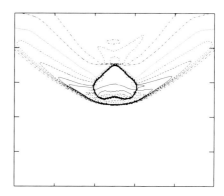

FIG. 2.6. Finding the right shape to enter the atmosphere using the characteristic function of the body (see Chapter 6 for shape parameterization issues). The final shape does not have the same regularity at the leading and trailing edges as the initial guess.

Another way is to extend the operators by zero inside S and take the characteristic function of Ω, χ, for unknown

$$\min_{\chi \in X_d} \int_D |\psi - \psi_d|^2 \; : \; -\nabla \cdot [\chi \nabla \psi] = 0, \quad \psi(1-\chi) = 0, \quad \psi|_{\partial\Omega} = \psi_d. \quad (2.18)$$

This last approach, suggested by Tartar [55] and Cea [20] has led to what is now referred as *numerical topological optimization*.

It may be difficult to work with the function χ. Then, following [3], the function χ can be defined through a smooth function η by $\chi(x) = \text{bool}(\eta(x) > 0)$ and in the algorithm we can work with a smooth η as in the level set methods.

Most existence results are obtained by considering minimizing sequences S^n, or T^n or χ^n and, in the case of our academic example, show that $\psi^n \to \psi$ for some ψ (resp. $T^n \to T$ or $\chi^n \to \chi$), and that the PDE is satisfied in the limit.

By using regularity results with respect to the domain Chenais [21] (see also [43] and [52]) showed that in the class of all S *uniformly Lipschitz*, problem (2.20) has a solution. However the solution could depend upon the Lipschitz constant.

Similarly Murat and Simon in [39] working with (2.13) showed that in the class of $T \in W^{1,\infty}$ uniformly, the solution exists.

However working with (2.18) generally leads to weaker results because if $\chi^n \to \chi$, χ may not be a characteristic function; this leads to a *relaxed problem*, namely (2.18) with

$$X_d = \{\chi \,:\, 0 \leq \chi(x) \leq 1\} \text{ instead of } \tilde{X}_d = \{\chi \,:\, \chi(x) = 0 \text{ or } 1\}. \quad (2.19)$$

These relaxed problems usually have a solution and it is often possible to show that if the solution is not in $\tilde{X}_d$ then it is the limit of a *composite domain* made of mixtures of smaller and smaller subdomains and holes [40].

In 2D and for Dirichlet problems like (2.20) there is a very elegant result due to Sverak [53] which shows that either there is no solution because the minimizing sequences converge to a composite domain or there is a regular solution; more precisely: *if a maximum number of connected components for the complement of Ω is imposed as an inequality constraint for the set of admissible domains then the solution exists.*

For fluids it is hard to imagine that any minimal drag geometry would be the limit of holes, layers of matter and fluids. Nevertheless in some cases the approach is quite powerful because it can answer *topological* questions which are embarrassing for the formulations (2.20) and (2.13) such as: is it better to have a long wing or two short wings for an airplane? Or is it better to have one pipe or two smaller pipes [16, 17] to move a fluid from left to right?

It is generally believed that control problems where the coefficients (here T) are the controls are numerically more difficult than shape optimization. Accordingly the second approach should be simpler as it involves a smaller parameter space.

Before proceeding we need the following preliminary result. In most cases only one part of the boundary Γ is optimized; we call this part S.

Proposition 2.3 *Consider a small perturbation S' of S given by*

$$S' = \{x + \lambda \alpha n : x \in S\},$$

where α is a function of $x(s) \in S$ (s is its curvilinear abscissa) and λ is a positive number destined to tend to zero. Denote $\Omega' = \Omega(S')$. Then for any f continuous in the neighborhood of S we have

$$\int_{\Omega'} f - \int_{\Omega} f = \lambda \int_S \alpha f + o(\lambda).$$

Proof

$$\int_{\Omega'} f - \int_{\Omega} f = \int_{\delta\Omega^+} f - \int_{\delta\Omega^-} f,$$

where

$$\delta\Omega^+ = \Omega' \backslash (\Omega' \cap \Omega), \qquad \delta\Omega^- = \Omega \backslash (\Omega' \cap \Omega).$$

In a neigborhood $\mathcal{O}$ of $x \in S$ a change of coordinates which changes n into $(0, ..., 0, 1)^T$ gives

$$\Omega \cap \mathcal{O} = \{(x_1, ..., x_{d-1}, y)^T \in \mathcal{O} : y < 0\}.$$

Then the contribution of $\mathcal{O}$ to the integral of f in $\delta\Omega^+$ is

$$\int_{S \cap \mathcal{O}} \left(\int_0^{\lambda\alpha} f(x_1, ..., x_{d-1}, x_d + y) dy \right) dx_1 ... dx_{d-1}.$$

Applying the mean value theorem to the integral in the middle gives, for some ξ (function of x)

$$\int_{S \cap \mathcal{O}} (\lambda \alpha f(x_1, ..., x_{d-1}, x_d + \xi) dy) \, dx_1 ... dx_{d-1}.$$

By continuity of f the results follow because the same argument can be used for $\delta\Omega^-$ with a change of sign.

If S has an angle not all variations S' can be defined by local variation on S. However the use of n for local variations is not mandatory: any vector field not parallel to S works. Similarly, the following can be proved [47]:

Proposition 2.4 *If $g \in H^1(S)$ and if R denotes the mean radius of curvature of S ($1/R = 1/R_1 + 1/R_2$ in 3D where R_1, R_2 are the principal radius of curvature) then*

$$\int_{S'} g - \int_S g = \lambda \int_S \alpha \left(\frac{\partial g}{\partial n} - \frac{g}{R} \right) + o(\lambda).$$

2.5.1 Sensitivity analysis for the nozzle problem

Let $D \subset \mathcal{R}^2$, $a \in \mathcal{R}$, $g \in H^1(\mathcal{R}^2)$ and $u_d \in L^2(D)$ and consider

$$\min_{\partial \Omega \in \mathcal{O}} \int_D |\nabla \phi - u_d|^2$$
$$\text{subject to}: \ -\Delta \phi = 0 \text{ in } \Omega, \ a\phi + \partial_n \phi = g \text{ on } \partial\Omega. \tag{2.20}$$

If $a = 0$ it is the potential flow formulation (2.10) and if $a \to \infty$ and $g = af$ it is the stream function formulation (2.11).

Let us assume that all shape variations are in the set of admissible shapes $\mathcal{O}$. Assume that some part of $\Gamma = \partial\Omega$ is fixed, the unknown part being called S. The variational formulation of (2.20) is

Find $\phi \in H^1(\Omega)$ such that
$$\int_\Omega \nabla \phi \cdot \nabla w + \int_\Gamma a\phi w = \int_\Gamma gw, \quad \forall w \in H^1(\Omega).$$

The Lagrangian of the problem is

$$L(\phi, w, S) = \int_D |\nabla \phi - u_d|^2 + \int_{\Omega(S)} \nabla \phi \cdot \nabla w + \int_\Gamma (a\phi w - gw),$$

and (2.20) is equivalent to the min-max problem

$$\min_{S,\phi} \max_v L(\phi, v, S).$$

Apply the min-max theorem and then derive the optimality condition to find that:
$$J'_S(S,\phi) = L'_S(\phi, v, S) \quad \text{at the solution } \phi, v \text{ of the min-max.}$$

Let us write that the solution is a saddle point of L.

$$\partial_\lambda L(\phi + \lambda\hat\phi, v, S)|_{\lambda=0} = 2\int_D (\nabla\phi - u_d) \cdot \nabla\hat\phi + \int_{\Omega(S)} \nabla\hat\phi \cdot \nabla v$$
$$+ \int_\Gamma a\hat\phi v = 0, \quad \forall \hat\phi$$

$$\partial_\lambda L(\phi, v + \lambda w, S)|_{\lambda=0} = \int_{\Omega(S)} \nabla\phi \cdot \nabla w + \int_\Gamma (a\phi w - gw) = 0 \quad \forall w.$$

According to the two propositions above, stationarity with respect to S is

$$\lim_{\lambda \to 0} \frac{1}{\lambda}[L(\phi, w, S') - L(\phi, w, S)] =$$
$$\int_S \alpha[\nabla\phi \cdot \nabla w + \partial_n(a\phi w - gw) - \frac{1}{R}(a\phi w - gw)] = 0,$$

and so we have shown that

Theorem 2.5 *The variation of J defined by (2.20) with respect to the shape deformation $S' = \{x + \alpha(x)n_S(x) : x \in S\}$ is*

$$\delta J \equiv J(S', \phi(S')) - J(S, \phi(S)) = \int_S \alpha[\nabla\phi \cdot \nabla v + \partial_n(a\phi v - gv)$$
$$- \frac{1}{R}(a\phi v - gv)] + o(\|\alpha\|),$$

where $v \in H^1(\Omega(S))$ is the solution of

$$\int_{\Omega(S)} \nabla v \cdot \nabla w + \int_\Gamma avw = -2\int_D (\nabla\phi - u_d) \cdot \nabla w, \quad \forall w \in H^1(\Omega(S)). \quad (2.21)$$

Corollary 2.6 *With Neumann conditions ($a = 0$)*

$$\delta J = -\int_S \alpha \left(\partial_s \phi \cdot \partial_s v + \frac{1}{R} vg\right) + o(\|\alpha\|),$$

and with Dirichlet conditions on S ($a \to \infty$)

$$\delta J = -\int_S \alpha \partial_n \phi \cdot \partial_n v + o(\|\alpha\|).$$

So, for problem (2.20) N iterations of the gradient method with Armijo rule is:
- S^0, $0 < \alpha < \beta < 1$;
- for n = 0 to N do
  * Solve the PDE (2.20) with $S = S^n$,
  * Solve the PDE (2.21) with $S = S^n$,
  * Compute $\gamma(x) = -[\nabla\phi \cdot \nabla v + \partial_n(a\phi v - gv) - \frac{1}{R}(a\phi v - gv)]$, $x \in S^n$,
  * Set $S(\rho) = \{x + \rho\gamma(x)\mathbf{n}(x), \ x \in S^n\}$,
  * Compute a ρ^n such that
  $$-\rho^n \beta \|\gamma\|^2 < J(\phi(S(\rho^n))) - J(S^n) < -\rho^n \alpha \|\gamma\|^2,$$
  * Set $S^{n+1} = S(\rho^n)$.
- done

This algorithm is still conceptual as the PDEs need be discretized and we must also be more precise on the norm for $\|\gamma\|$.

The generalization to the Navier-Stokes equation is given at the end of this chapter.

Alternatively the partial differential equation could be put into the cost function by penalty or penalty-duality and then the problem would be a more classical optimization of a functional; this is known as the one shot method [54]. Although it is simpler to solve the problem that way, it may not be the most efficient [56].

2.5.2 Numerical tests with freefem++

Freefem++ is a programming environment for solving partial differential equations by finite element methods. Programs are written in a language which is readible without explanation and follows the syntax of C++ for non-PDE-specific instructions. The program is open source and can be downloaded at www.freefem.org.

The following implements the gradient method with fixed step size for the academic nozzle problem treated above.

```
real xl = 5, L=0.3;
mesh th = square(30,30,[x,y*(0.2+x/xl)]);
func D=(x>0.4+L && x<0.6+L)*(y<0.1); real rho = 50;
func psid = 0.8*y;
fespace Vh(th,P1);
Vh psi,p,w,u,a=0.2+x/xl, gradJ;
problem streamf(psi,w) = int2d(th)(dx(psi)*dx(w) + dy(psi)*dy(w))
    + on(1,4,psi = y/0.2)
    + on(2,psi=y/(0.2+1.0/xl)) + on(3,psi=1);

problem adjoint(p,w) = int2d(th)(dx(p)*dx(w) + dy(p)*dy(w))
        - int2d(th)(D*(psi-psid)*w)+ on(1,2,3,4,p = 0);
for(int i=0;i<50;i++){
    streamf; adjoint;
```

```
    real J = int2d(th)(D*(psi-psid)^2)/2;
    gradJ = dx(psi)*dx(p) + dy(psi)*dy(p);
    a=a(x,0)-rho*gradJ(x,a(x,0))*x*(1-x);
    th = square(30,30,[x,y*a(x,0)]);
    plot(th,psi);
    if(i==0 || i==50)
      cout<<"J= "<<J<<" gradJ2= "<<int1d(th,3)(gradJ^2)<<endl;
}
```
Starting from a trapezoidal shape after 50 iterations the shape fixing ψ nearest to ψ_d is shown on the left in Fig. 2.7 for $D = (0.4, 0.6) \times (0, 0.1)$. When $D = (0.4, 0.6) \times (0, 10)$ there is no bounded solution. Furthermore oscillations develop on the free boundary. Figure 2.7 (center) shows them at iteration 5 (to obtain this, change the definition of D in the program above and change $\rho = 5$.

Oscillations can be controled by using Sobolev grandients for J. For this add in the program the definition of the operator "smooth" and change the definition of gradJ to the one below:

```
problem smooth(u,w) = int2d(th)(u*w+dx(u)*dx(w) + dy(u)*dy(w))
        - int2d(th)(w*(dx(psi)*dx(p) + dy(psi)*dy(p))) ;
...
smooth; gradJ = u;rho=500;
...
```

The result at iteration 5 is shown on Fig. 2.7 on the bottom.

2.6 Discretization with triangular elements

For discretization let us use the simplest, a finite element method of degree 1 on triangles. Unstructured meshes are better for OSD because they are easier to deform and adapt to any shape deformation.

More precisely, Ω is approximated by $\Omega_h = \cup_{k=1}^{n_T} T_k$ where the T_k are triangles such that:

- the vertices of $\partial\Omega_h$ are on $\partial\Omega$ and the corners of $\partial\Omega$ are vertices of $\partial\Omega_h$;
- $T_k \cap T_l$, $(k \neq l)$ is either a vertex or an entire edge or empty.

Triangulations are indexed by h the longest edge, and as $h \to 0$ no angle should go to zero or π.

The Sobolev space $H^1(\Omega)$ is approximated by continuous piecewise linear polynomials on the triangulation

$$H_h = \{w_h \in C^o(\bar{\Omega}_h) : w_h|_{T_k} \in P^1 \; \forall k\},$$

where $P^1 = P^1(T_k)$ is the space of linear polynomials on triangle T_k.

In variational form, a discretization of (2.20) is

$$\min_{\{(q^i)_{i=1}^{n_v} \in \mathcal{Q}\}} J(q^1, ..., q^{n_v}) = \int_D \|\nabla \phi_h - u_{dh}\|^2,$$

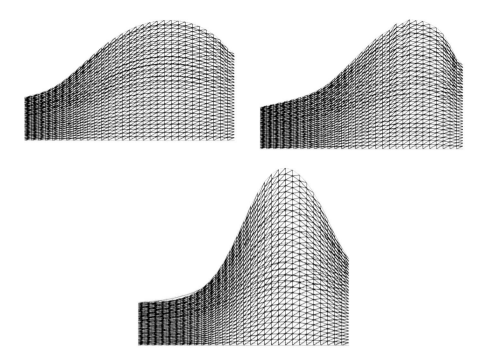

FIG. 2.7. *Solution of the academic problem with freefem++. The problem is to find the stream function as closed to a given stream function in a D. Left: $D = (0.4, 0.6) \times (0, 0.1)$, the solution after 50 iterations of a gradient method of the program 2.5.2; right and bottom $D = (0.4, 0.6) \times (0, 10)$, a case in which there is no bounded solution. Right corresponds to a gradient method without smoother and oscillations develop even at iteration 5 (shown). Bottom: the same with a gradient smoother (the Sobolev gradient of Chapter 6); there are no oscillations at iteration 5.*

subject to $\phi_h \in H_h$ solution of
$$\int_{\Omega_h} \nabla \phi_h \cdot \nabla w^j + \int_\Gamma a\phi_h w^j = \int_\Gamma g w^j \quad \forall j \in [1, ..., n_v] \qquad (2.22)$$

The dimension of H_h equals n_v the number of vertices q^i of the triangulation and every function ϕ_h belonging to H_h is completely determined by its values on the vertices $\phi_i = \phi_h(q^i)$.

The canonical basis of H_h is the set of so-called *hat functions* defined by

$$w^i \in H_h, \quad w^i(q^j) = \delta_{ij}.$$

Denoting by ϕ_i the coefficient of ϕ_h on that basis,

$$\phi_h(x) = \sum_1^{n_v} \phi_i w^i(x),$$

the PDE in variational form (2.22) yields a linear system for $\Phi = (\phi_i)$

$$A\Phi = F, \quad A_{ij} = \int_{\Omega_h} \nabla w^i \nabla w^j + \int_{\Gamma_h} aw^i w^j, \quad F_j = \int_{\Gamma_h} gw^j.$$

Hence the

Proposition 2.7 *In matrix form the discrete problem is:*
Find $q = (q^1, q^2, ..., q^{n_v})$ solution of

$$\min_{q \in \mathcal{Q}}\{J(q) = \Phi^T B \Phi - 2U \cdot \Phi \; : \; A(q)\Phi = F\}, \tag{2.23}$$

where

$$B_{ij} = \int_D \nabla w^i \nabla w^j, \quad U_j = \int_D u_d \cdot \nabla w^j.$$

Notice that B and U are independent of q if the triangulation is fixed inside D. For simplicity we shall assume that F does not depend on q either, i.e. that $g = 0$ on S.

The method applies also to Dirichlet conditions treated by penalty. However, in practice it is necessary for numerical quality to use a lumped quadrature in the integral of the Fourier term, or equivalently to apply $\phi = \phi_\Gamma$ at all points of Γ by another penalty

$$A_{ij} = \int_{\Omega_h} \nabla w^i \nabla w^j + p\delta_{ij} \mathbf{1}(q^i \in \Gamma_h), \quad F_j = p\phi_\Gamma(q^j)\mathbf{1}(q^j \in \Gamma_h),$$

where p is a large number and $\mathbf{1}(b)$ is 1 if b is true and zero otherwise.

We present below a computation of discrete gradients for a Neumann problem but the method applies also to Dirichlet conditions.

2.6.1 *Sensitivity of the discrete problem*

A straightforward calculus, first described by Marrocco and Pironneau [37], gives

$$\delta J = 2(B\Phi - U) \cdot \delta\Phi \quad \text{with} \quad A\delta\Phi = -(\delta A)\Phi.$$

Introducing Ψ a solution of $A^T \Psi = 2(B\Phi - U)$ leads to

$$\delta J = (A^T \Psi) \cdot \delta\Phi = \Psi^T A\delta\Phi = -\Psi \cdot ((\delta A)\Phi).$$

To evaluate δA we need three lemmas which relate all variations to δq, the vector of variation of vertex positions (i.e. each vertex q^i moves by δq^i). We define

$$\delta q_h(x) = \sum_1^{n_v} \delta q^i w^i(x), \quad \forall x \in \Omega_h,$$

and denote by Ω'_h the new domain.

Lemma 2.8
$$\delta w^j = -\nabla w^j \cdot \delta q_h + o(\|\delta q_h\|).$$

Lemma 2.9
$$\int_{\delta\Omega_h} f \equiv \int_{\Omega'_h} f - \int_{\Omega_h} f = \int_{\Omega_h} \nabla \cdot (f\delta q_h) + o(\|\delta q_h\|).$$

Lemma 2.10
$$\int_{\delta\Gamma_h} g \equiv \int_{\Gamma'_h} g - \int_{\Gamma_h} g = \int_{\Gamma_h} gt \cdot \partial_s \delta q_h + \int_{\Gamma_h} \delta q_h \nabla g + o(\|\delta q_h\|),$$

where ∂_s denotes the derivative with respect to the curvilinear abscissa and t the oriented tangent vector of Γ_h.

In these the integrals are sums of integrals on triangles or edges and so f and g can be piecewise discontinuous across elements or edges. Proofs for Lemmas 2.8 and 2.9 are in [48].

Proof of Lemma 2.10 Consider an edge $e_l = q^j - q^i$ and an integral on that edge

$$I_l = \|q^j - q^i\| \int_0^1 g(q^i + \lambda(q^j - q^i))d\lambda.$$

Then

$$\delta I_l = (\delta q^j - \delta q^i) \cdot (q^j - q^i)\|q^j - q^i\|^{-2} I_l$$
$$+ \|q^j - q^i\| \int_0^1 ((1-\lambda)\delta q^i + \lambda \delta q^j) \nabla g(q^i + \lambda(q^j - q^i))d\lambda + o(\delta q_h)$$
$$= \int_{\Gamma_h} gt \cdot \partial_s \delta q_h + \int_{\Gamma_h} \delta q_h \nabla g + o(\delta q_h).$$

Proposition 2.11
$$\delta J = \int_{\Omega_h} \nabla \psi_h^T (\mathbf{I}\nabla \cdot \delta q_h - \nabla \delta q_h - (\nabla \delta q_h)^T) \nabla \Phi_h$$
$$+ a \int_{\Gamma_h} \psi_h \cdot \Phi_h t^T (\nabla \delta q_h) t + o(\|\delta q_h\|),$$

where $\mathbf{I}$ is the identity matrix, t is the tangent vector to Γ_h, $\psi_h = \sum \Psi_i w^i$, and Ψ is solution of $A^T \Psi = 2(B\Phi - U)$.

Corollary 2.12 *The change of cost function due to inner nodes is $O(h^2)$ and*

$$\delta J = \int_\Gamma \nabla \psi_h \cdot \nabla p_h q_h \cdot n + a \int_{\Gamma_h} \psi_h \cdot \Phi_h t^T (\nabla \delta q_h) t + o(\|\delta q_h\|) + o(h).$$

Indeed consider a change of variable $x = Z_h(y)$, linearized at $Z_h(y) = y + z_h(y)$; it gives

$$\int_{\Omega_h} \nabla \psi_h \nabla p_h dx = \int_{Z_h^{-1}(\Omega)} (\nabla Z_h^{-1} \nabla \psi_h) \cdot (\nabla Z_h^{-1} \nabla p_h) \det \nabla Z_h) dy$$
$$\approx \int_{\Omega_h} \nabla \psi_h \cdot (I + \nabla z_h + \nabla z_h^T - \nabla \cdot z_h) \nabla p_h + \int_{Z_h^{-1}(\Omega_h) - \Omega_h} \nabla \psi_h \cdot \nabla p_h.$$

Therefore

$$\int_{\Omega_h} \nabla \psi_h \cdot (\nabla \cdot z_h - \nabla z_h - \nabla z_h^T) \nabla p_h \approx \int_{Z_h^{-1}(\Omega_h) - \Omega_h} \nabla \psi_h \cdot \nabla p_h$$
$$\approx \int_{\Gamma_h} \nabla \psi_h \cdot \nabla p_h (z_h \cdot n).$$

This corollary links the continuous and the discrete cases. In the continuous case we considered variations of the boundary only while in the discrete case we have to account for the variations of the inner vertices. The corollary shows that a change of position of an inner node has a second-order effect on the cost function compared with a change of position of a boundary node in the direction normal to the boundary.

Proof of Proposition 2.11
Now putting the pieces together

$$\delta \int_{\Omega_h} \nabla w^i \cdot \nabla w^j \simeq \int_{\delta\Omega_h} \nabla w^i \cdot \nabla w^j + \int_{\Omega_h} [\nabla \delta w^i \cdot \nabla w^j + \nabla w^i \cdot \nabla \delta w^j]$$
$$\simeq \int_{\Omega_h} [\nabla \cdot (\delta q_h \nabla w^i \cdot \nabla w^j) - \nabla(\nabla w^i \cdot \delta q_h) \cdot \nabla w^j$$
$$- \nabla w^i \cdot \nabla(\nabla w^j \cdot \delta q_h)],$$
$$\delta \int_{\Gamma_h} w^i \cdot w^j \simeq \int_{\delta\Gamma_h} w^i \cdot w^j + \int_{\Gamma_h} [\delta w^i \cdot w^j + w^i \cdot \delta w^j]$$
$$\simeq \int_{\Gamma_h} w^i \cdot w^j t \cdot \partial_s \delta q_h + \int_{\Gamma_h} \delta q_h \nabla (w^i \cdot w^j)$$
$$- \int_{\Gamma_h} [(\nabla w^i \cdot \delta q_h) \cdot w^j + (\nabla w^j \cdot \delta q_h) \cdot w^i].$$

Consequently an iterative process like the method of steepest descent to compute the optimal shape will move each vertex of the triangulation in the direction opposite to the partial derivative of J with respect to the vertex coordinates (e^k is the kth unit vector of R^d). This is the gradient method for (2.22):

- **Choose** an initial domain and a triangulation for it. Denote by $(q^1, ..., q^{n_v})$ the vertices of the triangulation.
- **for n = 1 to N** Replace q_k^i, $k = 1, .., d$ $i = 1, ..., n_v$ by

$$q_k^i := q_k^i - \rho [\int_{\Omega_h} \nabla \psi_h^T (\nabla \cdot (e^k w^i) - \nabla(e^k w^i) - (\nabla(e^k w^i))^T) \nabla \phi_h$$

$$+ a \int_{\Gamma_h} (\psi_h \cdot \phi_h) t^T \nabla(e^k w^i) t],$$

where ρ is computed by the Armijo rule.

2.7 Implementation and numerical issues

Computation of discrete derivatives of cost functional is, as we have seen, crafty work, reasonable only for simple problems.

Another difficulty is that for practical applications the definition of the optimization problem is often changed by the designer until a feasible problem is found: a cost function and constraint sets are tried, but the solution is found to violate certain unexpected constraints so the constraint set is changed; later multi-point optimization may be desired so the cost function and equations are changed, and each time the discrete gradients must be computed. *Automatic differentiation* is the cure but as we shall see it has its own difficulties.

Mesh deformation is also a big problem. After a few iterations the mesh is no longer feasible. Remeshing will induce interpolation errors of data defined on the old mesh. It may cause divergence of the optimization algorithm if done too often. *Automatic mesh adaptation* and motion is the cure but it must be done in the correct functional framework; this will also be explained in a later chapter.

Finally, boundary oscillations are also a frequent curse usually due to an incorrect choice of scalar product in the optimization algorithm. We will give some elements of answers in later chapters.

2.7.1 Independence from the cost function

After discretization a typical OSD problem is of the form

$$\min_{q \in \mathcal{Q}} J(q, \Phi) \; : \; A(q, \Phi) = 0,$$

where q is the set of vertices of the triangulation, $\mathcal{Q}$ the set of admissible positions for the vertices, A the state equation (usually nonlinear), and Φ the degrees of freedom of the state variables.

Sensitivity analysis gives

$$\delta J = J'_q \delta q + J'_\Phi \delta \Phi = (J'_q - \Psi^T A'_q)\delta q,$$

where Ψ is the solution of

$$A'^T_\Phi \Psi = J'_\Phi.$$

So J appears in two places and by its partial derivatives. This is a very useful, as we will see later, to set up incomplete gradients.

Finite differences These can be approximated by finite differences because:

$$\frac{\partial J}{\partial \Phi_j} \simeq \frac{J(q, \Phi + \delta \Phi_j) - J(q, \Phi)}{\delta \Phi_j}, \quad \frac{\partial J}{\partial q_k^j} \simeq \frac{J(q + \delta q_k^j e^k, \Phi) - J(q, \Phi)}{\delta q_k^j}.$$

The number of elementary computations is of order n_v. Indeed, if n_v is the number of mesh nodes, the calculation cost has the same complexity as the solution of a Laplace equation [8, 9]. So this computation can be considered as not too expensive. The method applied to compute A_Φ or A'_q, on the other hand, would be too expensive.

By this we have independence with respect to J. However, the results are sensitive to the values of $\delta\Phi_j$ and δq_k^j.

2.7.2 Addition of geometrical constraints

To add geometrical constraints is easy if we give a parameterized description of the domain and its triangulation.

If the boundary to optimize is described by r parameters α_j, we can define it by a curve (e.g. spline) defined by α_j and then generate the triangulation with vertices $\{q^i\}, i = 1, \ldots, n_v$ on the curve.

Since in this case only the parameters α_j move independently, we must compute the variation of J with respect to α_j. But

$$\frac{\partial J}{\partial \alpha_j} = \sum_{k,i} \frac{\partial J}{\partial q_i^k} \cdot \frac{\partial q_i^k}{\partial \alpha_j}, \quad i = 1, \ldots, n_v, \quad k = 1, 2.$$

Therefore, we must be able to compute $\frac{\partial q_i^k}{\partial \alpha_j}$ and this is done also by finite differences:

$$\frac{\partial q_i^k}{\partial \alpha_j} \simeq \frac{q_i^k(\alpha_j + \delta\alpha_j) - q_i^k(\alpha_j)}{\delta\alpha_j},$$

which is not computationally expensive.

One might think that everything could be computed by finite differences, even

$$\frac{\partial J}{\partial q_i^k} \simeq \frac{J(q_i^k + \delta q_i^k) - J(q_i^k)}{\delta q_i^k},$$

but this is far too expensive, since we have to solve the state equation every time we compute $J(q_i^k)$. Thus, the computational cost is $O(n_v^2)$ which is the cost of solution of n_v partial differential equations.

2.7.3 Automatic differentiation

Of course, finite difference approximations can be replaced by exact formal derivatives obtained with programs like Maple or Mathematica. However, this may not be so easy because J is usually an integral and the dependence on q is complex.

Then it has been noted that functions given by their computer programs are in principle easy to differentiate because every line of the program is easy to differentiate formally.

Efficient software such as (the list is not exhaustive):
Adol-C : http://www.math.tu-dresden.de/ adol-c/ [24],
Adifor: http://www.cs.rice.edu/ adifor/) [10],
Tapenade: http://www-sop.inria.fr/tropics/tapenade.html [50],

are now available. The adjoint and the linearized state equation can even be obtained automatically this way. We shall devote a chapter to this approach.

2.8 Optimal design for Navier-Stokes flows

2.8.1 *Optimal shape design for Stokes flows*

The drag and lift are the only forces at work in the absence of gravity. If the body is symmetric and its axis is aligned with the velocity at infinity then there is no lift and therefore we can equally well minimize the energy of the system, which for Stokes flow gives the following problem:

$$\min_{\Omega \in \mathcal{O}} J(\Omega) = \nu \int_\Omega |\nabla u|^2 \text{ subject to}$$
$$-\nu \Delta u + \nabla p = 0, \quad \nabla \cdot u = 0 \text{ in } \Omega \quad u|_S = 0, \quad u|_{\Gamma_\infty} = u_\infty.$$

An example of $\mathcal{O}$ is:

$$\mathcal{O} = \{\Omega, \ \partial\Omega = S \cup \Gamma_\infty, |\tilde{S}| = 1\},$$

where $\tilde{S}$ is the domain inside the closed boundary S and $|\tilde{S}|$ is its volume or area in 2D.

Sensitivity analysis is as before; let $\Omega' \in \mathcal{O}$ be a domain "near" Ω defined by its boundary $\Gamma' = \partial\Omega'$, with

$$\Gamma' = \{x + \alpha(x)n(x), \text{ with } \alpha = \text{ regular, small, } \forall x \in \Gamma = \partial\Omega\}.$$

Define also

$$\delta u = u(\Omega') - u(\Omega) \equiv u' - u,$$

while extending u by zero in $\tilde{S}$. Then

$$\delta J = \nu \delta \left(\int_\Omega |\nabla u|^2 \right) = \nu \int_{\delta\Omega} |\nabla u|^2 + 2\nu \int_\Omega \nabla \delta u : \nabla u + o(\delta\Omega, \delta u).$$

When ∇u is smooth, then

$$\nu \int_{\delta\Omega} |\nabla u|^2 = \nu \int_\Gamma \alpha |\nabla u|^2 + o(\|\alpha\|_{C^2}) = \nu \int_\Gamma \alpha |\partial_n u|^2 + o(\|\alpha\|_{C^2}).$$

Now $\delta u, \delta p$ satisfy

$$-\nu \Delta \delta u + \nabla \delta p = 0, \quad \nabla \cdot \delta u = 0 \text{ in } \Omega$$
$$\delta u|_{\Gamma_\infty} = 0, \quad \delta u|_S = -\alpha \partial_n u,$$

Indeed the only non obvious relation is the boundary condition on S. Now by a Taylor expansion

$$u'(x + \alpha n) = u'(x) + \alpha \partial_n u'|_{S'} + o(|\alpha|) = 0 \text{ since } u'|_{S'} = 0$$

Now $u|_S = 0$ so,
$$\delta u|_S = -\alpha \partial_n u|_S$$
Consequently ($A : B$ means $\sum_{ij} A_{ij} B_{ij}$)

$$\begin{aligned}
\nu \int_\Omega \nabla \delta u : \nabla u &= \nu \int_\Omega (-\Delta u) \cdot \delta u + \nu \int_\Gamma \partial_n u \cdot \delta u \\
&= \int_\Omega p \nabla \cdot \delta u - \int_\Gamma p \delta u \cdot n + \nu \int_\Gamma \partial_n u \cdot \delta u \\
&= \int_\Gamma (\nu \partial_n u - pn) \cdot \delta u \\
&= -\int_S \nu \alpha |\partial_n u|^2
\end{aligned}$$

because, if s denotes the tangent component,
$$n \cdot \partial_n u = -s \cdot \partial_s u = 0 \text{ on } \Gamma.$$

We have proved

Proposition 2.13 *The variation of J with respect to Ω is*
$$\delta J = -\nu \int_S \alpha |\partial_n u|^2 + o(\|\alpha\|).$$

Two consequences of this proposition are:
- If $\mathcal{O} = \{S : S \supset C\}$, as $|\partial_n u|^2 > 0$, then C is the solution (no change around C will decrease the drag in Stokes flow).
- If $\mathcal{O} = \{S : Vol \, \tilde{S} = 1\}$, then the object with minimum drag satisfies $\partial_n u \cdot s$ constant on S. Lighthill [46] showed that near the leading and the trailing edge the only possible axisymmetric flow which can achieve this condition must have conical tips of half angle equal to 60°. The method of steepest descent gives a shape close to the optimal shape after one iteration [46].

2.8.2 Optimal shape design for Navier-Stokes flows

A similar analysis can be done for the Navier-Stokes equation for incompressible flows. The result is as follows:

Proposition 2.14 *The minimum drag/energy problem with the Navier-Stokes equations is*

$$\min_{S \in \mathcal{O}} J(\Omega) = \nu \int_\Omega |\nabla u|^2 \quad \text{subject to}$$
$$-\nu \Delta u + \nabla p + u \nabla u = 0, \quad \nabla \cdot u = 0, \quad \text{in } \Omega,$$
$$u|_S = 0, \quad u|_{\Gamma_\infty} = u_\infty,$$

and with $\mathcal{O} = \{S : |\tilde{S}| = 1\}$, $\Gamma = \partial \Omega = S_\infty \cup \Gamma$. The variation of J with respect to normal shape variations α is

$$\delta J = \nu \int_S \alpha(\frac{\partial P}{\partial n} - \partial_n u) \cdot \partial_n u + o(\alpha),$$

where (P,q) is the solution of the linearized adjoint Navier-Stokes equation:

$$-\nu \Delta P + \nabla q - u\nabla P - (\nabla P)u = -2\nu \Delta u, \quad \nabla \cdot P = 0 \;\; in \;\; \Omega \;\; P|_\Gamma = 0.$$

Proof
Let us express the variation of $J(\Omega)$ in terms of the variation α of Ω. As for the Stokes problem,

$$\delta J = J(\Omega') - J(\Omega) = \nu \int_{\delta\Omega} |\nabla u|^2 + 2\nu \int_\Omega \nabla u \nabla \delta u + o(\delta u, \alpha),$$

but now the equation for δu is no longer self-adjoint

$$-\nu \Delta \delta u + \nabla \delta p + u \nabla \delta u + \delta u \nabla u = o(\delta u), \quad \nabla \cdot \delta u = 0,$$
$$\delta u|_{\Gamma_\infty} = 0, \quad \delta u|_S = -\alpha \partial_n u.$$

Multiply the equation for (P, q) by δu and integrate by parts

$$\int_\Omega \nu \nabla P : \nabla \delta u - \int_S \partial_n P \cdot \delta u - q\nabla \cdot \delta u + P\nabla \cdot (u \otimes \delta u + \delta u \otimes u)$$
$$= 2\int_\Omega \nu \nabla u : \nabla \delta u - 2\int_\Gamma \nu \partial_n u \cdot \delta u.$$

Then use the equation for δu multiplied by P and integrated on Ω

$$\int_\Omega \nu \nabla P : \nabla \delta u + P\nabla \cdot (u \otimes \delta u + \delta u \otimes u) = 0.$$

So

$$\delta J = \nu \int_{\delta\Omega} \alpha |\nabla u|^2 + \int_S \alpha(\partial_n P - 2\partial_n u) \cdot \partial_n u.$$

References

[1] Achdou, Y. (1992). Effect of a metallized coating on the reflection of an electromagnetic wave. *CRAS*, **314(3)**, 1-8.
[2] Achdou, Y. Valentin, F. and Pironneau, O. (1998). Wall laws for rough boundaries, *J. Comput. Phys.* **147**, 187-218.
[3] Allaire, G. (2007). *Conception Optimale de Structures*. Springer-SMAI, Paris.
[4] Allaire, G. Bonnetier, E. Frankfort, G. and Jouve, F. (1997) Shape optimization by the homogenization method, *Numerische Mathematik*, **76**, 27-68.
[5] Allaire, G. and Kohn, R.V. (1993), Optimal design for minimum weight and compliance in plane stress using extremal microstructures, *Europ. J. Mech. A/Solids*, **12(6)**, 839-878.

[6] Anagnostou, G. Ronquist, E. and Patera, A. (1992) A computational procedure for part design, *Comp. Methods Appl. Mech. Eng.* **20**, 257-270.

[7] Artola, M. and Cessenat, M. (1990). Propagation des ondes electromagnetiques dans un milieu composite, *CRAS*, **311(1)**, 77-82.

[8] Arumugam, G. (1989). *Optimum design et applications a la mecanique des fluides*, Thesis, University of Paris 6.

[9] Arumugam, G. and Pironneau, O. (1989). On the problems of riblets as a drag reduction device, *Optimal Control Appl. Meth.*, **10**, 332-354.

[10] Bischof, C. Carle, A. Corliss, G. Griewank, A. and Hovland, P. (1992) ADIFOR : Generating derivative codes from fortran programs, *Sci. Program*, **1(1)**, 11-29.

[11] Banichuk, N.V. (1990). *Introduction to Optimization of Structures*, Springer-Verlag, Berlin.

[12] Baron, A. (1998). *Optimisation multi-physique*, Thesis, University of Paris 6.

[13] Bensoe, Ph. (1995). *Optimization of Structural Topology, Shape, and Material*, Springer-Verlag, Berlin.

[14] Bensoussan, A. Lions, J.L. and Papanicolaou, G. (1978). *Asymptotic Analysis for Periodic Structures*, Studies in Mathematics and its Applications, Vol. 5, North-Holland, New York.

[15] Beux, F. and Dervieux, A. (1991). Exact-gradient shape optimization of a 2D Euler flow, *INRIA report*, **1540**.

[16] Borrval, T. and Petersson, J. (2003). Topological optimization in Stokes flow. *Int. J. Numer. Meth. Fluids*, **41(1)**, 77-107.

[17] Bangtsson, E. Noreland, D. and Berggren, M. (2002). *Shape optimization of an acoustic horn*. TR 2002-019 Uppsala university.

[18] Bucur, D. and Zolezio, J.P. (1995). Dimensional shape optimization under capacitance constraint, *J. Diff. Eqs.* **123(2)**, 504-522.

[19] Cea, J. (1986). *Conception optimale ou identification de forme: calcul rapide de la dérivée directionelle de la fonction coût*, Modélisation Math. Anal., AFCET, Dunod, Paris.

[20] Cea, J. Gioan, A. and Michel, J. (1973). Some results on domain identification, *Calcolo*, **3(4)**, 207-232.

[21] Chenais, D. (1987). Shape optimization in shell theory, *Eng. Opt.* **11**, 289-303.

[22] Choi, H. Moin, P. and Kim, J. (1992). *Turbulent drag reduction: studies of feedback control and flow over riblets*. CTR report **TF-55**.

[23] Dicesare, N. (2000). *Outils pour l'optimisation de forme et le contrôle optimal*, Thesis, University of Paris 6.

[24] Griewank, A. Juedes, D. and Utke, J. (1996) Algorithm 755: ADOL-C : a package for the automatic differentiation of algorithms written in C/ C++, *j-TOMS*, **22(2)**, 131-167.

[25] Gunzburger, M.D. (2002). *Perspectives in Flow Control and Optimization*. SIAM, Philadelphia.

[26] Haslinger, J. and Neittaanmäki, P. (1989). *Finite Element Approximations for Optimal Shape Design*, John Wiley, London.
[27] Haslinger, J. and Makinen, R.A.E. (2003). *Introduction to Shape Optimization*. Siam Advances in Design and Control, New York.
[28] Haug, E.J. and Cea, J. (1981). *Optimization of Distributed Parameter Structures* **I-II**, Sijthoff and Noordhoff, Paris.
[29] Herskowitz, J. (1992). An interior point technique for nonlinear optimization, *INRIA report*, **1808**.
[30] Isebe, D. Azerad, P. Bouchette, F. and Mohammadi, B. (2008). Optimal shape design of coastal structures minimizing short waves impact, *Coastal Eng.*, **55(1)**, 35-46.
[31] Isebe, D. Azerad, P. Bouchette, F. and Mohammadi, B. (2008). Shape optimization of geotextile tubes for sandy beach protection, *Inter. J. Numer. Meth. Eng.*, **25**, 475-491.
[32] Jameson, A. (1990). Automatic design of transonic airfoils to reduce the shock induced pressure drag, *Proc. 31st Israel Annual Conf. on Aviation and Aeronautics*.
[33] Kawohl, B. Pironneau, O. Tartar, L. and Zolesio, JP. (1998). *Optimal Shape Design*, Springer Lecture Notes in Mathematics, Berlin.
[34] Laporte, E. (1999). *Optimisation de forme pour écoulements instationnaires*, Thesis, Ecole Polytechnique.
[35] Lions, J.-L. (1968). *Contrôle Optimal des Systèmes Gouvernés par des Equations aux Dérivées Partielles*, Dunod, Paris.
[36] Mäkinen, R. (1990). Finite Element design sensitivity analysis for non linear potential problems, *Comm. Appl. Numer. Meth.* **6**, 343-350.
[37] Marrocco, A. and Pironneau, O. (1977). Optimum design with Lagrangian finite element, *Comp. Meth. Appl. Mech . Eng.* **15(3)**, 512545.
[38] Mohammadi, B. (1999). Contrôle d'instationnaritées en couplage fluide-structure. *Comptes rendus de l'Académie des Sciences*, **327(1)**, 115-118.
[39] Murat, F. and Simon, J. (1976). Etude de problèmes d'optimum design. *Proc. 7th IFIP Conf. Lecture Notes in Computer sciences*, **41**, 54-62.
[40] Murat, F. and Tartar, L. (1997). On the control of coefficients in partial differential equations. *Topics in the Mathematical Modelling of Composite Materials*, A. Cherkaev and R. Kohn eds. Birkhauser, Boston.
[41] Nae, C. (1998). Synthetic jets influence on NACA 0012 airfoil at high angles of attack, *American Institute of Aeronautics and Astronautics*, **98-4523**.
[42] Neittaanmäki, P. (1991). Computer aided optimal structural design. *Surv. Math. Ind.* **1**. 173-215.
[43] Neittaanmäki, P. and Stachurski, A. (1992). Solving some optimal control problems using the barrier penalty function method, *Appl. Math. Optim.* **25**, 127-149.
[44] Neittaanmaki, P. Sprekels, J. and Tiba, D. (2006). *Optimization of Elliptic Systems*. Springer, Berlin.

[45] Neittaanmaki, P. Rudnicki, M. and Savini, A. (1996). *Inverse Problems and Optimal Design in Electricity and Magnetism*. Oxford Science Publications.

[46] Pironneau, O. (1973). On optimal shapes for Stokes flow, *J. Fluid Mech.* **70(2)**, 331-340.

[47] Pironneau, O. (1984). *Optimal Shape Design for Elliptic Systems*, Springer, Berlin.

[48] Pironneau, O. (1983). *Finite Element in Fluids*, Masson-Wiley, Paris.

[49] Polak, E. (1997) *Optimization: Algorithms and Consistent Approximations*, Springer, New York.

[50] Rostaing, N. Dalmas, S. and Galligo, A. (1993). Automatic differentiation in Odyssee, *Tellus*, **45a(5)**, 558-568.

[51] Rousselet, B. (1983). Shape design sensitivity of a membrane, *J. Optim. Theory Appl.* **40(4)**, 595-623.

[52] Sokolowski, J. and Zolezio, J.P. (1991), *Introduction to Shape Optimization.* Springer Series in Computational Mathematics, Berlin.

[53] Sverak, A. (1992). On existence of solution for a class of optimal shape design problems, *CRAS*, **992**, 1-8.

[54] Ta'asan, S. and Kuruvila, G. (1992). Aerodynamic Design and optimization in one shot, *AIAA*, **92-0025**.

[55] Tartar, L. (1974). *Control Problems in the Coefficients of PDE*, Lecture notes in Economics and Math Systems. Springer, Berlin.

[56] Vossinis, A. (1995). Optimization algorithms for optimum shape design problems, Thesis university of Paris 6.

3
PARTIAL DIFFERENTIAL EQUATIONS FOR FLUIDS

3.1 Introduction

This chapter is devoted to a short description of the main state equations for fluid flows solved for the direct problem. As the design approach described throughout the book is universal, these equations can therefore be replaced by other partial differential equations. However, as we will see, knowledge of the physics of the problem and the accuracy of the solution are essential in shape design using control theory. In particular, we develop wall function modeling considered as reduced order models in incomplete sensitivity evaluation through the book.

3.2 The Navier-Stokes equations

Denote by Ω the region of space (R^3) occupied by the fluid. Denote by $(0, T^1)$ the time interval of interest. A Newtonian fluid is characterized by a density field $\rho(x,t)$, a velocity vector field $u(x,t)$, a pressure field $p(x,t)$, a temperature field $T(x,t)$ for all $(x,t) \in \Omega \times (0, T^1)$.

3.2.1 Conservation of mass

The variation of mass of fluid in $\mathcal{O}$ has to be equal to the mass flux across the boundaries of $\mathcal{O}$. So if n denotes the exterior normal to the boundary $\partial \mathcal{O}$ of $\mathcal{O}$,

$$\partial_t \int_{\mathcal{O}} \rho = - \int_{\partial \mathcal{O}} \rho u \cdot n,$$

By using the Stokes formula

$$\int_{\mathcal{O}} \nabla.(\rho u) = \int_{\partial \mathcal{O}} \rho u \cdot n,$$

and the fact that $\mathcal{O}$ is arbitrary, the *continuity equation* is obtained:

$$\partial_t \rho + \nabla.(\rho u) = 0. \tag{3.1}$$

3.2.2 Conservation of momentum

The forces on $\mathcal{O}$ are the external forces f (gravity for instance) and the force that the fluid outside $\mathcal{O}$ exercises, $\sigma n - pn$ per volume element, by definition of the stress tensor σ. Hence Newton's law, written for a volume element $\mathcal{O}$ of fluid gives

$$\int_{\mathcal{O}} \rho \frac{du}{dt} = \int_{\partial \mathcal{O}} (\sigma \mathbf{n} - pn).$$

Now

$$\frac{du}{dt}(x,t) = \lim_{\delta t \to 0} \frac{1}{\delta t}[u(x+u(x,t)\delta t, t+\delta t) - u(x,t)]$$
$$= \partial_t u + \sum_j u_j \partial_j u \equiv \partial_t u + u\nabla u,$$

$$\rho(\partial_t u + u\nabla u) + \nabla p - \nabla.\sigma = f.$$

By the continuity equation, this equation is equivalent to the *momentum equation*:

$$\partial_t(\rho u) + \nabla.(\rho u \otimes u) + \nabla.(p\mathbf{I} - \sigma) = f. \tag{3.2}$$

Newtonian flow To proceed further an hypothesis is needed to relate the stress tensor σ to u. For *Newtonian flows* σ is assumed linear with respect to the deformation tensor:

$$\sigma = \mu(\nabla u + \nabla u^T) + (\xi - \frac{2\mu}{3})\mathbf{I}\nabla.u. \tag{3.3}$$

The scalars μ and ξ are called the *first and second viscosities* of the fluid. For air and water the second viscosity ζ is very small, so $\xi = 0$ and the momentum equation becomes

$$\partial_t(\rho u) + \nabla.(\rho u \otimes u) + \nabla p - \nabla.[\mu(\nabla u + \nabla u^T) - \frac{2\mu}{3}\mathbf{I}\nabla.u] = f. \tag{3.4}$$

3.2.3 Conservation of energy and and the law of state

Conservation of energy is obtained by noting that the variation of the total energy in a volume element balances heat variation and the work of the forces $\mathcal{O}$. The energy $E(x,t)$ per unit mass in a volume element $\mathcal{O}$ is the sum of the internal energy e and the kinetic energy $u^2/2$. The work done by the forces is the integral over $\mathcal{O}$ of $u \cdot (f + \sigma - p\mathbf{I})n$. By definition of the temperature T, if there is no heat source (e.g. combustion) the amount of heat received (lost) is proportional to the flux of the temperature gradient, i.e. the integral on $\partial \mathcal{O}$ of $\kappa \nabla T.n$. The scalar κ is called the *thermal conductivity*. So the following equation is obtained

$$\frac{d}{dt}\int_{\mathcal{O}(t)} \rho E = \int_{\mathcal{O}} \{\partial_t \rho E + \nabla.[u\rho E]\}$$
$$= \int_{\mathcal{O}} u \cdot f + \int_{\partial \mathcal{O}} [u(\sigma - p\mathbf{I}) - \kappa \nabla T]n. \tag{3.5}$$

With the continuity equation and the Stokes formula it is transformed into

$$\partial_t[\rho E] + \nabla.(u[\rho E + p]) = \nabla.(u\sigma + \kappa \nabla T) + f \cdot u.$$

To close the system a definition for e is needed. For an *ideal fluid*, such as air and water in nonextreme situations, C_v and C_p being physical constants, we have

$$e = C_v T, \quad E = C_v T + \frac{u^2}{2}, \tag{3.6}$$

and the *equation of state*

$$\frac{p}{\rho} = RT, \tag{3.7}$$

where $\mathbf{R}$ is the ideal gas constant. With $\gamma = C_p/C_v = R/C_v + 1$, the above can be written as

$$e = \frac{p}{\rho(\gamma - 1)}. \tag{3.8}$$

With the definition of σ, the equation for E becomes what is usually referred to as *the energy equation*:

$$\partial_t \left[\rho \frac{u^2}{2} + \frac{p}{\gamma - 1}\right] + \nabla . \{u \left[\rho \frac{u^2}{2} + \frac{\gamma}{\gamma - 1} p\right]\} \tag{3.9}$$

$$= \nabla . \{\kappa \nabla T + \left[\mu(\nabla u + \nabla u^T) - \frac{2}{3}\mu \mathbf{I}\nabla . u\right] u\} + f \cdot u. \tag{3.10}$$

By introducing the *entropy*:

$$s \equiv \frac{R}{\gamma - 1} \log \frac{p}{\rho^\gamma}, \tag{3.11}$$

another form of the energy equation is the entropy equation:

$$\rho T (\partial_t s + u \nabla s) = \frac{\mu}{2}|\nabla u + \nabla u^T|^2 - \frac{2}{3}\mu|\nabla . u|^2 + \kappa \Delta T.$$

3.3 Inviscid flows

In many instances viscosity has a limited effect. If it is neglected, together with the temperature diffusion ($\kappa = 0, \eta = \xi = 0$), the equations for the fluid become the *Euler equations*:

$$\partial_t \rho + \nabla . (\rho u) = 0, \tag{3.12}$$

$$\partial_t \left[\rho \frac{u^2}{2} + \frac{p}{\gamma - 1}\right] + \nabla . \{u \left[\rho \frac{u^2}{2} + \frac{\gamma}{\gamma - 1} p\right]\} = f \cdot u. \tag{3.13}$$

Notice also that, in the absence of shocks, the equation for the entropy (3.11) becomes

$$\frac{\partial s}{\partial t} + u \nabla s = 0.$$

Hence, s is constant on the lines tangent at each point to u (stream-lines). In fact a stream-line is a solution of the equation :
$$x'(\tau) = u(x(\tau), \tau),$$
and so
$$\frac{d}{dt}s(x(t),t) = \frac{\partial s}{\partial x_i}\frac{\partial x_i}{\partial t} + \frac{\partial s}{\partial t} = \partial_t s + u\nabla s = 0.$$

If s is constant and equal to s^0, constant at time 0, and if s is also equal to s^0 on the part of Γ where $u \cdot n < 0$, then there is an analytical solution $s = s^0$.

Finally there remains a system of two equations with two unknowns, for *isentropic flows*
$$\partial_t \rho + \nabla.(\rho u) = 0, \qquad (3.14)$$
$$\rho(\partial_t u + u\nabla u) + \nabla p = f, \qquad (3.15)$$
where $p = C\rho^\gamma$ and $C = e^{s^0(\gamma-1)/R}$.

3.4 Incompressible flows

When the variations of ρ are small (water for example or air at low velocity) we can neglect its derivatives. Then the general equations become the incompressible *Navier-Stokes equations*
$$\nabla.u = 0, \qquad (3.16)$$
$$\partial_t u + u\nabla u + \nabla p - \nu\Delta u = f/\rho, \qquad (3.17)$$
with $\nu = \mu/\rho$ the *kinematic viscosity* and $p \to p/\rho$ the reduced pressure. If buoyancy effects are present in f, we need an equation for the temperature.

An equation for the temperature T and an analytic expression for ρ as a function of T can be obtained from the energy equation
$$\partial_t T + u\nabla T - \frac{\kappa}{\rho C_v}\Delta T = \frac{\nu}{2C_v}|\nabla u + \nabla u^T|^2. \qquad (3.18)$$

3.5 Potential flows

For suitable boundary conditions the solution of the Navier-Stokes equations can be irrotational
$$\nabla \times u = 0.$$
By the theorem of De Rham there exists then a *potential* function such that
$$u = \nabla\varphi.$$
Using the identities:

$$\Delta u = -\nabla \times \nabla \times u + \nabla(\nabla.u), \quad u\nabla u = -u \times (\nabla \times u) + \nabla(\frac{u^2}{2}), \quad (3.19)$$

when u is a solution of the incompressible Navier-Stokes equations (3.17) and $f = 0$, φ is a solution of the *Laplace equation*:

$$\Delta \varphi = 0 \quad (3.20)$$

and the *Bernoulli* equation, derived from (3.17), gives the pressure

$$p = k - \frac{1}{2}|\nabla \varphi|^2. \quad (3.21)$$

This type of flow is the simplest of all.

In the same way, with isentropic inviscid flow (3.15)

$$\partial_t \rho + \nabla \varphi \nabla \rho + \rho^0 \Delta \varphi = 0,$$

$$\nabla \left(\varphi_{,t} + \frac{1}{2}|\nabla \varphi|^2 + \gamma C \rho^{0\gamma-1} \rho \right) = 0.$$

If we neglect the convection term $\nabla \varphi \nabla \rho$ this system simplifies to a nonlinear *wave equation* :

$$\partial_{tt}\varphi - c\Delta \varphi + \frac{1}{2}\partial_t |\nabla \varphi|^2 = d(t), \quad (3.22)$$

where $c = \gamma C \rho^{0\gamma}$ is related to the velocity of the sound in the fluid.

Finally, we show that there are stationary potential solutions of Euler equations (3.13) with $f = 0$. Using (3.19) the equations can be rewritten as

$$-\rho u \times \nabla \times u + \rho \nabla \frac{u^2}{2} + \nabla p = 0.$$

Taking the scalar product with u, we obtain:

$$u \cdot \left[\rho \nabla \frac{u^2}{2} + \nabla p \right] = 0.$$

Also, the pressure being given by (3.15)

$$u \cdot \left(\rho \nabla \frac{u^2}{2} + \rho^{\gamma-1} C\gamma \nabla \rho \right) = 0.$$

or equivalently

$$u\rho.\nabla \left(\frac{u^2}{2} + C\frac{\gamma}{\gamma-1}\rho^{\gamma-1} \right) = 0.$$

So the quantity between the parentheses is constant along the stream lines; that is we have

$$\rho = \rho^0 \left(k - \frac{u^2}{2} \right)^{1/(\gamma-1)}.$$

Indeed the solution of the PDE $u\nabla \xi = 0$, in the absence of shocks, is ξ constant on the streamlines. If it is constant upstream (on the inflow part of the boundary where $u \cdot n < 0$), and if there are no closed streamlines then ξ is constant everywhere. Thus if ρ^0 and k are constant upstream then $\nabla \times u$ is parallel to u and, at least in two dimensions, this implies that u derives from a potential. The *transonic potential flow equation* is then obtained

$$\nabla \cdot [(k - |\nabla \varphi|^2)^{1/(\gamma-1)} \nabla \varphi] = 0. \tag{3.23}$$

The time-dependent version of this equation is obtained from

$$u = \nabla \varphi, \quad \rho = \rho^0 (k - (\frac{u^2}{2} + \partial_t \varphi)^{1/(\gamma-1)}), \quad \partial_t \rho + \nabla \cdot (\rho u) = 0.$$

3.6 Turbulence modeling

In this section we consider the Navier-Stokes equations for incompressible flows (3.17).

3.6.1 The Reynolds number

Let us rewrite (3.17) in non-dimensional form. Let U be a characteristic velocity of the flow under study (for example, one of the non homogeneous boundary conditions). Let L be a characteristic length (for example, the diameter of Ω) and T_1 a characteristic time (which is a priori equal to L/U). Let us put

$$u' = \frac{u}{U}; \quad x' = \frac{x}{L}; \quad t' = \frac{t}{T_1}.$$

Then (3.17) can be rewritten as

$$\nabla_{x'} \cdot u' = 0, \quad \frac{L}{T_1 U} \partial_{t'} u' + u' \nabla_{x'} u' + \frac{1}{U^2} \nabla_{x'} p - \frac{\nu}{LU} \Delta_{x'} u' = f \frac{L}{U^2}.$$

So, if we put $T_1 = L/U$, $p' = p/U^2$, $\nu' = \nu/LU$, then the equations are the same but with prime variables. The inverse of ν' is called the *Reynolds number*:

$$Re = \frac{UL}{\nu}.$$

3.6.2 Reynolds equations

Consider (3.17) with random initial data, $u^0 = \bar{u}^0 + u'^0$, where $\bar{u}$ stands for the expected value. Taking the expected value of the Navier-Stokes equations leads to

$$\nabla \cdot \bar{u} = 0, \tag{3.24}$$

$$\partial_t \bar{u} + \nabla \cdot \overline{(\bar{u} + u') \otimes (\bar{u} + u')} + \nabla \bar{p} - \nu \Delta \bar{u} = \bar{f}, \tag{3.25}$$

which is also

$$\nabla \cdot \bar{u} = 0, \quad R = -\overline{u' \otimes u'}, \tag{3.26}$$

$$\partial_t \bar{u} + \nabla \cdot (\bar{u} \otimes \bar{u}) + \nabla \bar{p} - \nu \Delta \bar{u} = \bar{f} + \nabla \cdot R. \tag{3.27}$$

3.6.3 The $k - \varepsilon$ model

Reynolds hypothesis is that the turbulence in the flow is a local function of $\nabla \bar{u} + \nabla \bar{u}^T$:

$$R(x,t) = R(\nabla \bar{u}(x,t) + \nabla \bar{u}^T(x,t)). \tag{3.28}$$

If the turbulence is locally isotropic at scales smaller than those described by the model (3.27) and if the Reynolds hypothesis holds, then it is reasonable to express R on the basis formed with the powers of $\nabla \bar{u} + \nabla \bar{u}^T$ and to relate the coefficients to the two turbulent quantities used by Kolmogorov to characterize homogeneous turbulence: the kinetic energy of the small scales k and their rate of viscous energy dissipation ε:

$$k = \frac{1}{2}\overline{|u'|^2}, \qquad \varepsilon = \frac{\nu}{2}\overline{|\nabla u' + \nabla u'^T|^2}. \tag{3.29}$$

For two-dimensional mean flows (for some $\alpha(x,t)$)

$$R = \nu_T(\nabla \bar{u} + \nabla \bar{u}^T) + \alpha I, \quad \nu_T = c_\mu \frac{k^2}{\varepsilon}, \tag{3.30}$$

and k and ε are modeled by

$$\partial_t k + \bar{u}\nabla k - \frac{c_\mu}{2}\frac{k^2}{\varepsilon}|\nabla \bar{u} + \nabla \bar{u}^T|^2 - \nabla.(c_\mu\frac{k^2}{\varepsilon}\nabla k) + \varepsilon = 0, \tag{3.31}$$

$$\partial_t \varepsilon + \bar{u}\nabla \varepsilon - \frac{c_1}{2}k|\nabla \bar{u} + \nabla \bar{u}^T|^2 - \nabla.(c_\varepsilon\frac{k^2}{\varepsilon}\nabla \varepsilon) + c_2\frac{\varepsilon^2}{k} = 0, \tag{3.32}$$

with $c_\mu = 0.09, c_1 = 0.126, c_2 = 1.92, c_\varepsilon = 0.07$.

The model is derived heuristically from the Navier-Stokes equations with the following hypotheses:

- Frame invariance and 2D mean flow, ν_t a polynomial function of k, ε.
- u'^2 and $|\nabla \times u'|^2$ are passive scalars when convected by $\bar{u} + u'$.
- Ergodicity allows statistical averages to be replaced by space averages.
- Local isotropy of the turbulence at the level of small scales.
- A Reynolds hypothesis for $\overline{\nabla \times u' \otimes \nabla \times u'}$.
- A closure hypothesis: $\overline{|\nabla \times \nabla \times u'|^2} = c_2\, \varepsilon^2/k$.

The constants $c_\mu, c_\varepsilon, c_1, c_2$ are chosen so that the model reproduces

- the decay in time of homogeneous turbulence;
- the measurements in shear layers in local equilibrium;
- the log law in boundary layers.

The model is *not valid near solid walls* because the turbulence is not isotropic so the near wall boundary layers are removed from the computational domain. An adjustable artificial boundary is placed parallel to the walls Γ at a distance $\delta(x,t) \in [10, 100]\nu/u_\tau$.

A possible set of boundary conditions is then

$\overline{u}, k, \varepsilon$ given initially everywhere;
$\overline{u}, k, \varepsilon$ given on the inflow boundaries at all t;
$\nu_T \partial_n \overline{u}, \nu_T \partial_n k, \nu_T \partial_n \varepsilon$ given on outflow boundaries at all t

$$u \cdot n = 0, \quad \frac{\overline{u} \cdot s}{\sqrt{\nu |\partial_n \overline{u}|}} - \frac{1}{\chi} \log\left(\delta\sqrt{\frac{1}{\nu}|\partial_n \overline{u}|}\right) + \beta = 0 \quad \text{on } \Gamma + \delta;$$

$$k|_{\Gamma+\delta} = |\nu \partial_n(\overline{u} \cdot s)| c_\mu^{-\frac{1}{2}}, \quad \varepsilon|_{\Gamma+\delta} = \frac{1}{\chi\delta}|\nu \partial_n(\overline{u} \cdot s)|^{\frac{3}{2}};$$

where $\chi = 0.41, \beta = 5.5$ for smooth walls, n, s are the normal and tangent to the wall, and δ is a function such that at each point of $\Gamma + \delta$,

$$10\sqrt{\nu/|\partial_n(\overline{u} \cdot s)|} \leq \delta \leq 100\sqrt{\nu/|\partial_n(\overline{u} \cdot s)|}.$$

These wall functions are classical and are only valid for regions where the flow is attached and the turbulence fully developed in the boundary layer. Below, we show how to derive generalized wall functions valid up to the wall and also valid for separated and unsteady flows.

3.7 Equations for compressible flows in conservation form

For compressible flows, let us consider the conservation form of the Navier-Stokes equations with the $k - \varepsilon$ model. As before, the model is derived by splitting the variables into mean and fluctuating parts and use Reynolds averages for the density and pressure and Favre averages for other variables. The non dimensionalized Reynolds averaged equations are closed by an appropriate modeling [7], and we have:

$$\partial_t \rho + \nabla \cdot (\rho u) = 0, \tag{3.33}$$

$$\partial_t(\rho u) + \nabla \cdot (\rho u \otimes u) + \nabla(p + \frac{2}{3}\rho k) = \nabla \cdot ((\mu + \mu_t)S), \tag{3.34}$$

$$\partial_t(\rho E) + \nabla \cdot ((\rho E + p + \frac{5}{3}\rho k)u) = \nabla \cdot ((\mu + \mu_t)Su) + \nabla((\chi + \chi_t)\nabla T), \tag{3.35}$$

with

$$\chi = \frac{\gamma\mu}{Pr}, \quad \chi_t = \frac{\gamma\mu_t}{Pr_t}, \tag{3.36}$$

$$\gamma = 1.4, \quad Pr = 0.72 \quad \text{and} \quad Pr_t = 0.9, \tag{3.37}$$

where μ and μ_t are the inverse of the laminar and turbulent Reynolds numbers. In what follows, we call them viscosities. The laminar viscosity μ is given by Sutherland's law:

$$\mu = \mu_\infty \left(\frac{T}{T_\infty}\right)^{1.5} \left(\frac{T_\infty + 110.4}{T + 110.4}\right), \tag{3.38}$$

where f_∞ denotes a reference quantity for f or its value at infinity if the flow is uniform there and

$$\mathbf{S} = \nabla u + \nabla u^T - \frac{2}{3}\nabla \cdot u\, I,$$

is the deformation tensor.

We consider the state equation for a perfect gas:

$$p = (\gamma - 1)\rho T.$$

Experience shows that almost everywhere $\rho k \ll p$, and we therefore drop the turbulent energy contributions in terms with first-order derivative (the hyperbolic part). This also improves the numerical stability, reducing the coupling between the equations [7].

The $k - \varepsilon$ model [3] we use is classical; it is an extension to compressible flows of the incompressible version [11]:

$$\partial_t \rho k + \nabla.(\rho u k) - \nabla((\mu + \mu_t)\nabla k) = S_k, \qquad (3.39)$$

and

$$\partial_t \rho \varepsilon + \nabla.(\rho u \varepsilon) - \nabla((\mu + c_\varepsilon \mu_t)\nabla \varepsilon) = S_\varepsilon. \qquad (3.40)$$

The right-hand sides of (3.39) and (3.40) contain the production and the destruction terms for ρk and $\rho \varepsilon$:

$$S_k = \mu_t P - \frac{2}{3}\rho k \nabla.u - \rho\varepsilon, \qquad (3.41)$$

$$S_\varepsilon = c_1 \rho k P - \frac{2c_1}{3c_\mu}\rho\varepsilon\nabla.u - c_2\rho\frac{\varepsilon^2}{k}. \qquad (3.42)$$

The eddy viscosity is given by:

$$\mu_t = c_\mu \rho \frac{k^2}{\varepsilon}. \qquad (3.43)$$

The constant c_μ, c_1, c_2 and c_ε are respectively $0.09, 0.1296, 11/6, 1/1.4245$ and $P = S : \nabla u$. The constant c_2 and c_ε are different from their original values of 1.92 and $1/1.3$. The constant c_2 is adjusted to reproduce the decay of k in isotropic turbulence. With $u = 0, \rho = \rho_\infty, T = T_\infty$ the model gives

$$k = k_0 \left(1 + (c_2 - 1)\frac{\varepsilon_0}{k_0}t\right)^{\frac{-1}{c_2 - 1}}. \qquad (3.44)$$

The experimental results of Comte-Bellot [1] give a decay of k in $t^{-1.2}$ and this fixes $c_2 = 11/6$ while $c_2 = 1.92$ leads to a decay in $t^{-1.087}$ and therefore to an overestimation of k.

This has also been reported in [9], where the author managed to compute the right recirculating bubble length for the backward step problem using the standard $k - \varepsilon$ model with this new value $c_2 = 11/6$, wall laws and $c_\varepsilon = 1/1.3$.

Finally, the compatibility relation between the $k - \varepsilon$ constants which comes from the requirement of a logarithmic velocity profile in the boundary layer [7] gives the c_ε constant:

$$c_\varepsilon = \frac{1}{\kappa^2 \sqrt{c_\mu}}(c_2 c_\mu - c_1) = \frac{1}{1.423}, \quad \kappa = 0.41,$$

to be compared to the classical value of $c_\varepsilon = 1/1.3$.

3.7.1 Boundary and initial conditions

The previous system of Navier-Stokes and $k - \varepsilon$ equations is well posed in the small, as the mathematicians say, meaning that the solution exists for a small time interval at least, with the following set of boundary conditions.

Inflow and outflow The idea is to avoid boundary layers such that all second-order derivatives are removed and that the remaining system (Euler-$k - \varepsilon$ model) is a system of conservation laws no longer coupled (as we dropped the turbulent contributions to first-order derivative terms). Inflow and outflow boundary conditions are of characteristic types. Roughly the idea is to impose the value of a variable if the corresponding wave enters the domain following the sign of the corresponding eigenvalue (in 3D):

$$\lambda_1 = u.n + c, \lambda_{2,3,4} = u.n, \lambda_5 = u.n - c, \lambda_{6,7} = u.n, \quad (3.45)$$

where n is the unit outward normal. However, as the system cannot be fully diagonalized, we use the following approach [8]. Along these boundaries the fluxes are split into positive and negative parts following the sign of the eigenvalues of the Jacobian A of the convective operator F.

$$\int_{\Gamma_\infty} F.n d\sigma = \int_{\Gamma_\infty} (A^+ W_{in} + A^- W_\infty).n d\sigma, \quad (3.46)$$

where W_{in} is the computed (internal) value at the previous iteration and W_∞ the external value, given by the flow.

Symmetry Here again the idea is to avoid boundary layers. We drop terms with second-order derivatives and the slipping boundary condition ($u.n = 0$) is imposed in weak form.

Solid walls The physical boundary condition is a no-slip boundary condition for the velocity ($u = 0$) and for the temperature, either an adiabatic condition ($\partial_n T = 0$) or an isothermal condition ($T = T_\Gamma$). However, the $k - \varepsilon$ model above is not valid [1] near walls because the turbulence is not isotropic at small scales. In the wall laws approach the near-wall region is removed from the computational domain and the previous conditions are replaced by by Fourier conditions of the type

$$u_\delta.n = 0, \quad u_\delta.t = f_1(\partial_n u_\delta, \partial_n T_\delta), \quad T_\delta = f_2(\partial_n u_\delta, \partial_n T_\delta) \quad (3.47)$$

for isothermal walls. This will be described in more detail later.

Initial conditions For external as well as internal flows, the initial flow is taken uniform with small values for k_0 and ε_0 (basically $10^{-6}|u_0|$). We take the same value for k and ε leading to a large turbulent time scale $k/\varepsilon = 1$ which characterizes a laminar flow:

$$u = u_0, \quad \rho = \rho_0, \quad T = T_0, \quad k = k_0, \quad \varepsilon = \varepsilon_0. \qquad (3.48)$$

Internal flow simulations often also require given profiles for some quantities. This is prescribed on the corresponding boundaries during the simulation.

3.8 Wall laws

The general idea in wall laws is to remove the stiff part from boundary layers, replacing the classical no-slip boundary condition by a more sophisticated relation between the variables and their derivatives. We introduce a constant quantity called friction velocity from:

$$\rho_w u_\tau^2 = (\mu + \mu_t)\frac{\partial u}{\partial y}\Big|_\delta = \mu \frac{\partial u}{\partial y}\Big|_0, \qquad (3.49)$$

where w means at walls and δ at a distance δ from the real wall. Using u_τ we introduce a local Reynolds number:

$$y^+ = \frac{\rho_w y u_\tau}{\mu_w}. \qquad (3.50)$$

The aim is now to express the behavior of $u^+ = u/u_\tau$ in term of y^+ which means that the analysis will be independent of the Reynolds number .

In this section we describe our approach to wall laws. We also give an extension to high-speed separated flows with adiabatic and isothermal walls. The ingredients are:

- global wall laws: numerically valid up to the wall (i.e. $\forall y^+ \geq 0$);
- weak formulation: pressure effects are taken into account in the boundary integrals which come from integrations by parts;
- small δ in wall laws: this means that the computational domain should not be too far from the wall;
- fine meshes: in the sense that the computational mesh should be fine enough for the numerical results to become mesh independent;
- compressible extension: laws valid for a large range of Mach number.

An important and interesting feature of wall laws is that they are compatible with an explicit time integration scheme, something which is not so real with low-Reynolds corrections.

3.8.1 *Generalized wall functions for u*

The first level in the modeling for wall laws is to consider flows attached (i.e. without separations) on adiabatic walls (i.e . $\frac{\partial T}{\partial n} = 0$). We are looking for laws valid up to the wall (i.e. valid $\forall y^+$). We consider the following approximated momentum equation in near-wall regions (x and y denote the local tangential and normal directions):

$$\partial_y((\mu + \mu_t)\partial_y u) = 0, \tag{3.51}$$

where

$$\mu_t = \kappa\sqrt{\rho\rho_w}yu_\tau(1 - e^{-y^+/70}), \quad \text{with } y^+ = \frac{\rho_w u_\tau y}{\mu_w}, \tag{3.52}$$

is a classical expression for the eddy viscosity valid up to the wall. Equation (3.51) means that the shear stress along y is constant. u_τ is a constant called the friction velocity and is defined by:

$$u_\tau = \left(\frac{(\mu + \mu_t)}{\rho_w}\frac{\partial u}{\partial y}\right)^{1/2} = \text{constant}, \tag{3.53}$$

where the subscript w means at the wall.

High-Reynolds regions In high-Reynolds regions the eddy viscosity dominates the laminar one and this leads to the log law with $\mu_t = \kappa\sqrt{\rho\rho_w}yu_\tau$,

$$\frac{\partial u}{\partial y} = \sqrt{\frac{\rho_w}{\rho}}\frac{u_\tau}{\kappa y}, \quad u = u_\tau\sqrt{\frac{\rho_w}{\rho}}\left(\frac{1}{\kappa}\log(y) + C\right), \tag{3.54}$$

provided that $\frac{\partial \rho}{\partial y} \ll \frac{\partial u}{\partial y}$ which is acceptable because $\frac{\partial \rho}{\partial y} \sim 0$ and $\frac{\partial T}{\partial y} = 0$ as the wall is adiabatic. Therefore $\frac{\partial \rho}{\partial y} \sim 0$.

We can see that at this level, there is no explicit presence of the Reynolds number. The dependency with respect to the Reynolds number is in the constant C. To have a universal expression, we write

$$u = u_\tau\sqrt{\frac{\rho_w}{\rho}}\left(\frac{1}{\kappa}\log(\frac{yu_\tau\rho_w}{\mu_w}) + \beta\right), \tag{3.55}$$

where $\beta = -\log(u_\tau\rho_w/\mu_w) + C)$ is found to have a universal value of about 5 for incompressible flows [2]:

$$u^+ = \frac{u}{u_\tau} = \sqrt{\frac{\rho_w}{\rho}}\left(\frac{1}{\kappa}\log(y^+) + 5\right). \tag{3.56}$$

Note that we always use wall values to reduce y-dependency as much as possible. This is important for numerical implementation.

Low-Reynolds regions In low-Reynolds regions, (3.52) is negligible and (3.51) gives a linear behavior for u vanishing at the walls:

$$\rho_w u_\tau^2 = \mu\frac{\partial u}{\partial y} \sim \mu\frac{u}{y}. \tag{3.57}$$

In other words, we have:

$$u^+ = \frac{u}{u_\tau} = \frac{yu_\tau\rho_w}{\mu_w} = y^+. \tag{3.58}$$

General expression To have a general expression, we define the friction velocity u_τ as a solution of

$$u = u_\tau \sqrt{\frac{\rho_w}{\rho}} f(u_\tau). \tag{3.59}$$

where f is such that $w = u_\tau \sqrt{\frac{\rho_w}{\rho}} f(u_\tau)$ is solution of (3.51) and (3.52). The wall-function therefore is not known explicitly and depends on the density distribution. A hierarchy of laws can be obtained therefore by taking into account compressibility effects starting from low-speed laws (see Section 3.10). Our aim during this development is to provide laws easy to implement for unstructured meshes. For low-speed flows, where density variations are supposed negligible, a satisfactory choice for f is the nonlinear Reichardt function f_r defined by :

$$f_r(y^+) = 2.5 \log(1 + \kappa y^+) + 7.8 \left(1 - e^{-y^+/11} - \frac{y^+}{11} e^{-0.33 y^+}\right). \tag{3.60}$$

This expression fits both the linear and logarithmic velocity profiles and also its behavior in the buffer region.

3.8.2 Wall function for the temperature

Consider the viscous part of the energy equation written in the boundary layer (i.e. $\partial_x \ll \partial_y$):

$$\frac{\partial}{\partial y}\left(u(\mu + \mu_t)\frac{\partial u}{\partial y}\right) + \frac{\partial}{\partial y}\left((\chi + \chi_t)\frac{\partial T}{\partial y}\right) = 0.$$

When we integrate this equation between the fictitious wall ($y = \delta$) and the real one ($y = 0$), we obtain:

$$(\chi + \chi_t)\frac{\partial T}{\partial y}|_\delta - \chi \frac{\partial T}{\partial y}|_0 = u(\mu + \mu_t)\frac{\partial u}{\partial y}|_0 - u(\mu + \mu_t)\frac{\partial u}{\partial y}|_\delta. \tag{3.61}$$

So, thanks to $\frac{\partial T}{\partial y}|_0 = 0$ and $u|_0 = 0$:

$$(\chi + \chi_t)\frac{\partial T}{\partial y}|_\delta + u(\mu + \mu_t)\frac{\partial u}{\partial y}|_\delta = 0. \tag{3.62}$$

Therefore, in the adiabatic case, there is no term for the energy equation to account for [5, 6] while for isothermal walls we need to close (3.63) providing an expression for either the first term in the left-hand side or for the right-hand side (see Section 3.10):

$$(\chi + \chi_t)\frac{\partial T}{\partial y}|_\delta + u(\mu + \mu_t)\frac{\partial u}{\partial y}|_\delta = \chi \frac{\partial T}{\partial y}|_0. \tag{3.63}$$

As a consequence, to evaluate the heat transfer at the wall, we have to use:

$$C_h = \frac{\chi \partial_y T|_0}{\rho_\infty u_\infty^3 \gamma} = \frac{(\chi + \chi_t)\partial_y T|_\delta + u \rho_w u_\tau^2}{\rho_\infty u_\infty^3 \gamma}.$$

This is important as industrial codes usually do the post-processing in a separate level than computation and the fluxes are not communicated between the two modules. In other words, with these codes, when using wall functions as well as with low-Reynolds models, only the first term is present in a heat flux evaluation above. This might also explain some of the reported weakness of wall functions for heat transfer

In separation and recirculation areas u and u_τ needed by our wall laws are small. As a consequence, this leads to an underestimation of the heat flux. In these areas, by dimension arguments, we choose the local velocity scale to be:

$$u = c_\mu^{-3/4}\sqrt{k}, \qquad (3.64)$$

and redefine, the friction flux by:

$$(\mu + \mu_t)\frac{\partial u}{\partial y} = c_\mu^{-3/4}(\mu + \mu_t)\frac{\partial \sqrt{k}}{\partial y}. \qquad (3.65)$$

3.8.3 k and ε

Once u_τ is computed, k and ε are set to:

$$k = \frac{u_\tau^2}{\sqrt{c_\mu}}\alpha, \qquad \varepsilon = \frac{k^{\frac{3}{2}}}{l_\varepsilon}, \qquad (3.66)$$

where $\alpha = \min(1,(\frac{y^+}{20})^2)$ reproduces the behavior of k when δ tends to zero (δ is the distance of the fictitious computational domain from the solid wall). The distance δ is given a priori and is kept constant during the computation. l_ε is a length scale containing the damping effects in the near-wall regions.

$$l_\varepsilon = \kappa c_\mu^{-3/4} y \left(1 - \exp(\frac{-y^+}{2\kappa c_\mu^{-3/4}})\right). \qquad (3.67)$$

Again, here the limitation is for separation points where the friction velocity goes to zero while a high level for k would be expected.

3.9 Generalization of wall functions

To extend the domain of validity of wall functions, the analytical functions above can be upgraded to account for compressibility effects, separation and unsteadiness. Beyond simulation, this is important for incomplete low-complexity sensitivity evaluation in shape optimization as we will see.

3.9.1 *Pressure correction*

This is an attempt to take into account the pressure gradient and convection effects in the classical wall laws.

To account for pressure and convection effects, f_c is added to the Reichardt equation (3.60):

$$f(u_\tau) = f_r(u_\tau) + f_c(u_\tau), \qquad (3.68)$$

f_c is a new contribution when pressure and convection effects exist.

The simplified momentum equation (3.51) is enhanced to

$$\partial_y((\mu + \mu_t)\partial_y u) = \partial_x p + \partial_x(\rho u^2) + \partial_y(\rho u v), \tag{3.69}$$

where μ_t is given by (3.52). Suppose that the right-hand side of equation (3.69) is known and constant close to the wall:

$$C = \partial_x p + \partial_x(\rho u^2) + \partial_y(\rho u v). \tag{3.70}$$

We can then integrate this equation in y, not exactly unfortunately, but only after a first-order development in y of the exponential near $y = 0$. The equation is written in terms of y^+:

$$\partial_{y^+}\left((1 + \kappa y^+(1 - e^{y^+/70}))\partial_{y^+} u\right) = C\mu u_\tau^{-2} = C', \tag{3.71}$$

so,

$$\partial_{y^+}((1 + y^{+2}\frac{\kappa}{70})\partial_{y^+}) = C', \tag{3.72}$$

hence,

$$\partial_{y^+} u = \frac{C'y^+}{1 + \frac{\kappa}{70}y^{+2}} + \frac{A}{1 + \frac{\kappa}{70}y^{+2}}. \tag{3.73}$$

After a second integration and using the boundary conditions (i.e. at the wall and at δ) we have:

$$u = \frac{35C\mu}{\kappa u_\tau^2} \log\left(1 + \frac{\kappa}{70}(y^+)^2\right). \tag{3.74}$$

On the other hand, the equation is also easily integrated when $\mu_t = \kappa \rho y u_\tau$. Hence the corrections are respectively

$$f_c(y^+) = \left(\frac{35C\nu}{\kappa u_\tau^3}\right) \log\left(1 + \frac{\kappa}{70}(y^+)^2\right) \quad \text{when} \quad y^+ \ll 70, \tag{3.75}$$

and

$$f_c(y^+) = \frac{C\delta}{\kappa u_\tau^2} \quad \text{elsewhere.} \tag{3.76}$$

These expressions intersect at $y^+ = 5.26$. This is in contradiction with the limit $y^+ = 70$ given by (3.52). However, as we would like the correction to perturb weakly the Reichardt law, we define the correction as the minimum of the two expressions. Hence, we apply (3.75) for $y^+ \leq 5.26$ and (3.76) for $y^+ \geq 5.26$. Of course, this correction vanishes with C and we recover the Reichardt law.

3.9.2 Corrections on adiabatic walls for compressible flows

Note that the previous wall laws are valid for incompressible flows. We therefore need to introduce some corrections to take into account the compressible feature of the flow. By now, ∞ will denote reference quantities and e will refer to the nearest local value outside boundary layer. We need to account for density variations in (3.59) and for the fact that Reichardt law has been suggested for low-speed flows.

3.9.3 Prescribing ρ_w

Let us define the recovery factor [2]:

$$r = \frac{T_f - T_e}{T_{ie} - T_e},$$

where T_f is called the friction temperature and T_{ie} is given by:

$$T_{ie} = T_e \left(1 + \frac{\gamma - 1}{2} M_e^2\right).$$

For turbulent flows, it is admitted that $r = Pr^{1/3}$ [2]. We obtain

$$T_f = T_e \left(1 + Pr^{\frac{1}{3}} \frac{\gamma - 1}{2} M_e^2\right).$$

In the adiabatic case, the wall temperature is the friction temperature T_f [2] (i.e. $T_w = T_f$).

To solve (3.51)-(3.52) in the adiabatic compressible case, we have to provide μ_w and ρ_w. The viscosity at the wall μ_w is obtained from the Sutherland law:

$$\mu_w = \mu_e \left(\frac{T_w}{T_e}\right)^{1/2} \frac{1 + 110.4/T_e}{1 + 110.4/T_w}. \tag{3.77}$$

For the second quantity, we use the Crocco relation [2]:

$$T = T_w + (T_{ie} - T_w) \frac{u}{u_e} - (T_{ie} - T_e) \left(\frac{u}{u_\infty}\right)^2.$$

As a consequence, we have:

$$\frac{T}{T_w} = 1 + \left[\left(1 + \frac{\gamma - 1}{2} M_e^2\right) \frac{T_e}{T_w} - 1\right] \frac{u}{u_e} - \frac{\gamma - 1}{2} M_e^2 \frac{T_e}{T_w} \left(\frac{u}{u_e}\right)^2.$$

We suppose the static pressure constant in the normal direction (i.e. $\partial_y p = 0$), therefore, from the perfect gas law, we obtain:

$$\frac{\rho_w}{\rho} = \frac{T}{T_w}.$$

As a consequence, we evaluate ρ_w thanks to:

$$\rho_w = \rho \left[1 + \left[\left(1 + \frac{\gamma - 1}{2} M_e^2\right) \frac{T_e}{T_w} - 1\right] \frac{u}{u_e} - \frac{\gamma - 1}{2} M_e^2 \frac{T_e}{T_w} \left(\frac{u}{u_e}\right)^2\right]. \tag{3.78}$$

However, the numerical implementation of (3.77) and (3.78) is not straightforward as we do not know the e denoted values (i.e. u_e, T_e, M_e) on unstructured meshes. We therefore choose to use only quantities known for any unstructured meshes: at the fictitious wall or at inflow. In particular, M_e is replaced by

$$M_\delta = \sqrt{\frac{u^2 + v^2}{\frac{\gamma P}{\rho}}}.$$

More precisely, knowing $(\rho_\delta, T_\delta, M_\delta, u_\infty, T_w)$ we find ρ_w by:

$$\rho_w = \rho_\delta \left[1 + \left[\left(1 + \frac{\gamma - 1}{2} M_\delta^2\right) \frac{T_\delta}{T_w} - 1 \right] \frac{u_\delta}{u_\infty} - \frac{\gamma - 1}{2} M_\delta^2 \frac{T_\delta}{T_w} \left(\frac{u_\delta}{u_\infty}\right)^2 \right]. \tag{3.79}$$

3.9.4 Correction for the Reichardt law

The next step is to introduce a correction for the Reichardt law. We can use one of the following approaches.

Using mixing length formula Following Cousteix [2], express the turbulent tension thanks to the mixing-length formula for high-Reynolds region ($\kappa y \partial_y u = u_\tau$):

$$\rho_w u_\tau^2 = \rho \kappa^2 y^2 (\partial_y u)^2,$$

so that

$$\partial_y u = \sqrt{\frac{\rho_w}{\rho}} \frac{u_\tau}{\kappa y}.$$

Express ρ_w/ρ thanks to the Crocco law and obtain:

$$\frac{\partial u}{\partial y} = \frac{u_\tau}{\kappa y} \sqrt{1 + b \frac{u}{u_\infty} - a^2 \left(\frac{u}{u_\infty}\right)^2}, \tag{3.80}$$

with

$$a^2 = \frac{\gamma - 1}{2} M_\delta^2 \frac{T_\delta}{T_w} \quad \text{and} \quad b = (1 + \frac{\gamma - 1}{2} M_\delta^2) \frac{T_\delta}{T_w} - 1$$

. The weakness of this approach is that it is not valid up to the wall. A global correction needs a global mixing-length formula as starting point using (3.59):

$$\partial_y u = u_\tau \left(\partial_y f(y^+) \sqrt{\frac{\rho_w}{\rho}} + f(y^+) \partial_y \left(\sqrt{\frac{\rho_w}{\rho}}\right) \right), \tag{3.81}$$

which is hardly computable.

A similar approach comes from Van Driest [2, 10], with the log-law relation as starting point in (3.83). Unfortunately, this also leads to a relation only valid for high-Reynolds region.

Restart from (3.51)-(3.52) To avoid the difficulty above, we would like to restart from our boundary layer system for u:

$$\left(\mu + \sqrt{\rho\rho_w}\kappa u_\tau y(1 - e^{-y^+/70})\right)\frac{\partial u}{\partial y} = \rho_w u_\tau^2. \qquad (3.82)$$

Now, suppose that the Reichardt law is obtained after integration of

$$u_\tau = \partial_y u \left(\mu + \kappa y(1 - e^{-y^+/70})\right). \qquad (3.83)$$

First consider the case $y^+ > 100$ and drop the laminar viscosity. Hence, replacing u_τ by (3.83) in the left-hand side of (3.82), leads to:

$$\frac{\partial u}{\partial y} = \sqrt[4]{\frac{\rho_w}{\rho}}\frac{u_\tau}{\kappa y(1 - e^{-y^+/70})}. \qquad (3.84)$$

The Crocco law links the density and temperature and (3.84) becomes:

$$\frac{\partial u}{\partial y} = \sqrt[4]{1 + b\frac{u}{u_\infty} - a^2\left(\frac{u}{u_\infty}\right)^2}\frac{u_\tau}{\kappa y(1 - e^{-y^+/70})}, \qquad (3.85)$$

The integration of relation (3.85) is not possible. At this level, we use the following approximation:

$$\left(\frac{1}{a}\left(\arcsin\frac{2a^2 u/u_\infty - b}{\sqrt{b^2 + 4a^2}} + \arcsin\frac{b}{(b^2 + 4a^2)^{1/2}}\right)\right)^{1/4} = u_\tau f_r(y^+). \qquad (3.86)$$

If laminar viscosity dominates the eddy one ($y^+ < 5$),

$$\rho_w u_\tau^2 = \mu \partial_y u. \qquad (3.87)$$

For $5 < y^+ < 100$, we use a linear interpolation between the two expressions above.

Numerical experiences have shown better behavior for this second approach.

3.10 Wall functions for isothermal walls

For isothermal walls ($T_w = T_{\text{given}}$), we have to provide a law for the temperature, as we did for the velocity. In weak form, we only need a law for the thermal stress $\chi\frac{\partial T}{\partial y}$.

The Reynolds relation A first attempt to evaluate $\chi\partial_y T$ is to use the classical Reynolds relation between heat and friction coefficients [2]:

$$C_h = \frac{s\,C_f}{2} = \frac{1.24}{2}C_f = 0.62 C_f, \qquad (3.88)$$

so, we have:
$$-\frac{\chi \partial_y T}{\rho u^3 \gamma} = C_h = 0.62 C_f = 1.24 \frac{\rho_w u_\tau^2}{\rho u^2}.$$

Note that, for the isothermal case, as $(\mu + \mu_t)\partial_y u = \rho_w u_\tau^2$, (3.61) leads to:
$$(\chi + \chi_t)\partial_y T \mid_\delta + u \rho_w u_\tau^2 = \chi \partial_y T.$$

So, we have [5]:
$$(\chi + \chi_t)\partial_y T \mid_\delta + u \rho_w u_\tau^2 = -1.24 \rho_w u_\tau^2 \gamma u. \tag{3.89}$$

Again, in the definition of the friction and heat coefficients, we use local values instead of reference ones. The wall density is obtained through Crocco's law (3.79).

Crocco's method A more general approach consists in linearizing the local Crocco law for the temperature to obtain the normal temperature slope. We have:
$$\frac{T}{T_w} = 1 + \left[\left(1 + \frac{\gamma-1}{2}M^2\right)\frac{T}{T_w} - 1\right]u - \frac{\gamma-1}{2}M^2 \frac{T}{T_w} u^2,$$

and, we obtain:
$$\frac{\partial T}{\partial y} = \left(\frac{\gamma-1}{2} 2M \frac{\partial M}{\partial y} T + \left(1 + \frac{\gamma-1}{2}M^2\right)\frac{\partial T}{\partial y}\right) u \tag{3.90}$$
$$+ [(1 + \frac{\gamma-1}{2}M^2)T - T_w]\frac{\partial u}{\partial y} \tag{3.91}$$
$$- \frac{\gamma-1}{2}(2MT\frac{\partial M}{\partial y} + M^2 \frac{\partial T}{\partial y}) u^2 \tag{3.92}$$
$$- \frac{\gamma-1}{2} M^2 T 2u \frac{\partial u}{\partial y}. \tag{3.93}$$

Moreover, by definition, $M^2 = u^2 \rho / \gamma p$. As a consequence:
$$\frac{\partial M}{\partial T} = \sqrt{\frac{\rho}{\gamma p}} \frac{\partial u}{\partial y} + \frac{u}{\sqrt{\gamma p}} \frac{1}{2\sqrt{\rho}} \frac{\partial \rho}{\partial y},$$

because the static pressure is supposed to be constant in the normal direction.

Finally, we have to express $\partial_y \rho$. Thanks to the perfect gas law and as the static pressure is constant in the normal direction, we find:
$$\frac{\partial p}{\partial y} = 0 = (\gamma-1)\left(\rho \frac{\partial T}{\partial y} + T \frac{\partial \rho}{\partial y}\right),$$

so
$$\frac{\partial \rho}{\partial y} = -\frac{\rho}{T}\frac{\partial T}{\partial y}.$$

As a consequence, we have:

$$\frac{\partial M}{\partial T} = \sqrt{\frac{\rho}{\gamma p}} \frac{\partial u}{\partial y} - \frac{M}{2T} \frac{\partial T}{\partial y}.$$

Reported in (3.93), we find:

$$\frac{\partial T}{\partial y} = \frac{1.5(\gamma-1)M^2 T + T - T_w - 2(\gamma-1)M^2 T u}{1-u} \frac{\partial u}{\partial y}$$

where $\partial_y u$ is given by (3.49):

$$\frac{\partial u}{\partial y} = \frac{\rho_w u_\tau^2}{\mu + \mu_t}.$$

At last, The temperature slope prescribed by Crocco's law is:

$$(\chi + \chi_t) \frac{\partial T}{\partial y}\bigg|_\delta = \frac{\chi + \chi_t}{\mu + \mu_t} \frac{(\gamma-1)M^2 T(1.5 - 2u) + T - T_w}{1-u} \rho_w u_\tau^2 \qquad (3.94)$$

where all the values are local ones. This expression is valid up to the wall and can be used for $\delta = 0$.

References

[1] Comte-Bellot, G. and Corsin, S. (1971). Simple Eulerian time-correlation of full and narrow-band velocity signals in grid-generated isotropic turbulence, *J. Fluid Mech.*, **48**, 273-337.

[2] Cousteix, J. (1990). *Turbulence et Couche Limite*, Cepadues, Paris.

[3] Launder, B.E. and Spalding, D.B. (1972). *Mathematical Models of Turbulence*, Academic Press, London.

[4] Mohammadi, B. (1992). Complex turbulent compressible flows computation with a two-layer approach, *Int. J. Num. Meth. for Fluids*, **15**, 747-771.

[5] Mohammadi, B. and Puigt, G. (2000). Generalized wall functions for high-speed separated flows over adiabatic and isothermal walls, *Int. J. Comp. Fluid Dynamics*, **14**, 20-41.

[6] Mohammadi, B. and Puigt, G. (2006). Wall functions in computational fluid mechanics, *Computers & Fluids*, **35(10)**, 1108-1115.

[7] Mohammadi, B. and Pironneau, O. (1994). *Analysis of the k-epsilon Turbulence Model*, John Wiley, London.

[8] Steger, J. and Warming, R.F. (1983). Flux vector splitting for the inviscid gas dynamic with applications to finite-difference methods, *J. Comp. Phys.* **40**, 263-293.

[9] Thangam, S. (1991). Analysis of two-equation turbulence models for recirculating flows, *ICASE report* **91-61**.

[10] Van Driest, E.R. (1951). Turbulent boundary layers in compressible fluids, *J. Aeronaut. Sci.*, **18**, 145-160.

[11] Vandromme, D. (1983). Contribution à la modélisation et la prédiction d'écoulements turbulents à masse volumique variable, *Thesis*, University of Lille.

4
SOME NUMERICAL METHODS FOR FLUIDS

4.1 Introduction

In this chapter we present the numerical methods used in this book to solve the equations for the flows. It is not an exhaustive presentation of computational fluid dynamics and since we are interested in optimization and control the chosen methods have been selected for their simplicity and easy treatment of general boundary conditions on complex geometries.

One important aspect is the implementation of wall laws for general separated and unsteady compressible or incompressible flows.

We present a mixed finite volume / finite element solver for compressible flows. The explicit character of this solver has influenced some of our later choice. This explicit approach is efficient enough on wall functions and appropriate for time-dependent flows. Today, it is accepted that to capture the flow features in large eddy simulations (LES) or unsteady Reynolds averaged Navier-Stokes (RANS) flows, the time step has to be similar to or smaller than the Courant-Fredriech-Levy stability condition (CFL) of explicit solvers. It is nice to notice that the CFL stability conditions on meshes needed for simulations with wall laws also resolves the large eddies of the flow in LES solvers.

Our choice of solver for the incompressible Navier-Stokes equations is explicit for the velocity field and implicit for the pressure. Our aim is to keep the solver as simple as possible and yet capable of an unstructured mesh with nonisotropic adaptivity. This way automatic differentiation (AD) of programs will be easier to apply.

At the end of this chapter, some simulations of complex flows, usually known to be difficult, demonstrate the abilities of both solvers.

4.2 Numerical methods for compressible flows

Here we present a solver for the Navier-Stokes equations in conservation form based on a finite volume Galerkin method and the Roe [27] flux for the approximation of the advection part. Of course, this is not the only choice possible and several other fluxes are available in the literature [13,33]. We begin with a short review of the possibilities.

4.2.1 *Flux schemes and upwinded schemes*

There are two major classes of scheme: flux difference splitting (FDS) like that of Roe and flux vector splitting (FVS) like in the Stegger and Warming scheme used here to implement general characteristic-compatible boundary conditions [28].

FDS schemes are more accurate for boundary layer computation as they capture stationary contacts, smoothed by FVS schemes. On the other hand, FVS schemes are more robust and cheap. An alternative family is the hybrid upwind schemes (HUS) of Coquel [6] where a correction is introduced to the FVS scheme to capture stationary contact. Our experience is that accuracy requirement makes FDS schemes more suitable even if at the expense of a somewhat longer computing time.

There is another major approach through streamline upwind, Petrov Galerkin (SUPG) approximations, and Galerkin least squares [19–21]. These are purely Galerkin schemes as the same finite element space is used for advection and diffusion and where the numerical dissipation is introduced by additional terms.

Each of these approaches has advantages and drawbacks. Upwind schemes for systems are usually designed for one-dimensional systems and their extension to multidimensional configurations is not straightforward, while there is usually no general artificial dissipation operator for systems, but when it exists there is usually no major difficulty for a generalization to higher dimensions.

4.2.2 A FEM-FVM discretization

We shall use a spatial discretization of the Navier-Stokes equations based on a finite volume Galerkin formulation. We use a Roe [27] Riemann solver for the convective part of the equations together with MUSCL reconstruction with Van Albada [30] type limiters. However, the limiters are only used in the presence of shocks. The viscous terms are treated using a Galerkin finite element method on linear triangular (tetrahedral) elements.

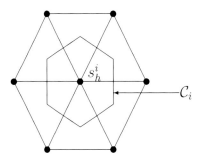

Consider the bidimensional Navier-Stokes equations in conservation form:

$$\frac{\partial W}{\partial t} + \nabla.(F(W) - N(W)) = S(W), \tag{4.1}$$

where $W = (\rho, \rho u, \rho v, \rho E, \rho k, \rho \varepsilon)^T$ is the vector of conservation variables, F and N are the convective and diffusive operators, and S contains the right-hand sides of the turbulence model.

Denote by $\{q_i\}_1^N$ the vertices of the triangulation. Let $\Omega_h = \cup_j T_j$ be a discretization by triangles of the computational domain Ω and let $\Omega_h = \cup_i C_i$ be its partition into cells; the cells boundaries are made by the lines which join a vertex and a middle point of the opposite edge in a triangle.

Let V_h be the set of continuous piecewise linear functions on our triangulation. We can associate to each $w_h \in V_h$, a w'_h piecewise constant function on cells by

$$w'_h|_{C_i} = \frac{1}{|C_i|} \int_{C_i} w_h. \tag{4.2}$$

Conversely, knowing w'_h piecewise constant, w_h is obtained as the piecewise linear continuous function which takes values $w_i \equiv w_h(q^i) = w'_h|_{C_i}$.

We use a Petrov Galerkin discretization of the weak formulation of (4.1). Find $W_h \in V_h^6$ such that, $\forall \Phi_h$

$$\int_{\Omega_h} \frac{\partial W_h}{\partial t} \Phi'_h - \int_{\Omega_h} (F_h - N_h)(W_h) \nabla(\Phi'_h) \tag{4.3}$$

$$+ \int_{\partial \Omega_h} (F_h - N_h) \cdot n \Phi'_h = \int_{\Omega_h} S_h \Phi_h. \tag{4.4}$$

Let Φ'_i be the characteristic function of C_i and let Φ_i be its corresponding function in V_h^6; with an explicit time discretization we obtain

$$|C_i| \left(\frac{W_i^{n+1} - W_i^n}{\delta t} \right) + \int_{\partial C_i} F_d(W^n) \cdot n = \text{RHS}, \tag{4.5}$$

$$\text{RHS} = - \int_{\Omega_h} N(W^n) \nabla(\Phi_i) + \int_{\partial \Omega} N(W^n) \cdot n \Phi_i. \tag{4.6}$$

Here, $F_d(W_h^n) = F(W_{\partial \Omega})$ on $\partial C_i \cap \partial \Omega$ and elsewhere F_d is a piecewise constant upwind approximation of $F(W)$ satisfying

$$\int_{\partial C_i} F_d \cdot n = \sum_{j \nu} \Psi(W'|_{C_i}, W'|_{C_j}) \int_{\partial C_i \cap C_j} n, \tag{4.7}$$

where Ψ is a numerical flux function satisfying, among other things, $\Psi(W,W) = F(W)$.

4.2.3 Approximation of the convection fluxes

To keep classical notation, for each segment (i,j), we denote the state at i by left and at j by right. We then define Roe's mean values for the velocity and total enthalpy (i.e. $H = |u|^2/2 + c^2/(\gamma - 1)$) using

$$u^R = \frac{u_l \sqrt{\rho_l} + u_r \sqrt{\rho_r}}{\sqrt{\rho_l} + \sqrt{\rho_r}}, \quad H^R = \frac{H_l \sqrt{\rho_l} + H_r \sqrt{\rho_r}}{\sqrt{\rho_l} + \sqrt{\rho_r}}, \tag{4.8}$$

from which we deduce the local speed of sound $c^R = ((\gamma - 1)(H^R - |u^R|^2/2))^{1/2}$. Using these, we can define the Jacobian matrix of F (i.e. $\partial F/\partial W$) for Roe's mean values [27]. Hence, we take for Φ the Roe flux

$$\Phi_{\text{Roe}}(W_l, W_r) = \frac{1}{2}(F(W_l) + F(W_r)) - \left|\frac{\partial F}{\partial W}(W^R)\right|(W^r - W^l). \qquad (4.9)$$

The absolute value is applied on the eigenvalue λ_j of the Jacobian [16]:

$$\left|\frac{\partial F}{\partial W}(W^R)\right| = \sum_j |\lambda_j| R_j L_j, \qquad (4.10)$$

where R_j and L_j are the right and left eigenvectors corresponding to λ_j.

4.2.4 Accuracy improvement

To improve the accuracy, we redefine the left and right states used in the previous scheme using Taylor expansions. This is called higher-order reconstruction. This is a fundamental difference between finite volumes and finite elements. In the first case, the function space degree is low (here P^0) and we use reconstruction to improve the accuracy while in the latter case, we try to increase the accuracy by changing the functional space, going for instance from P^1 to P^2.

Spatial accuracy is improved using geometrical state reconstruction [10] involving a combination of upwind and centered gradients. More precisely, let ∇W_i be an approximation of the gradient of W at node i. We define the following quantities on the segment $\vec{ij} = q^j - q^i$

$$W_{ij} = W_i + 0.5 \text{Lim}(\beta(\nabla W)_i \vec{ij}, (1-\beta)(W_i - W_j)), \qquad (4.11)$$

and

$$W_{ji} = W_j - 0.5 \text{Lim}(\beta(\nabla W)_j \vec{ij}, (1-\beta)(W_j - W_i)), \qquad (4.12)$$

with Lim being a Van Albada type limiter [30]:

$$\text{Lim}(a, b) = 0.5(1 + \text{sign}(ab)) \frac{(a^2 + \alpha)b + (b^2 + \alpha)a}{a^2 + b^2 + 2\alpha} \qquad (4.13)$$

with $0 < \alpha \ll 1$ and β a positive constant containing the amount of upwinding $\beta \in [0, 1]$ (here $\beta = 2/3$). Now, the second-order accuracy in space is obtained by replacing W_i' and W_j' in (4.7) by W_{ij} and W_{ji}.

Developed for structured meshes, these techniques have been successfully extended to unstructured meshes in the past [10].

4.2.5 Positivity

This approach does not guarantee the positivity of ρk and $\rho \varepsilon$, therefore the convective fluxes corresponding to the turbulence equations are computed using the scheme proposed in [22] for positivity preservation of chemical species.

More precisely, once the density fluxes are known the turbulent convective fluxes are computed by

$$\int_{\partial(C_i \cap C_j)} \rho u k.n = k_{ij} \int_{\partial(C_i \cap C_j)} \rho u \cdot n$$

$$k_{ij} = k_i \ (\text{resp.} k_j) \quad \text{if} \quad \int_{\partial(C_i \cap C_j)} \rho u \cdot n > 0 \quad (\text{resp.} < 0).$$

In [24, 26] we introduced two intermediate variables (ϕ, θ) forming a stable first-order time-dependent system in the absence of viscosity with a positive solution:

$$\varphi = k^a \varepsilon^b, \quad \theta = \frac{k}{\varepsilon}. \tag{4.14}$$

The $k - \varepsilon$ equations presented in Chapter 2 are written symbolically in terms of total derivatives D_t as

$$D_t k = S_k, \quad D_t \varepsilon = S_\varepsilon. \tag{4.15}$$

Differentiating (4.14), we have:

$$D_t \varphi = k^{a-1} \varepsilon^{b-1} (a\varepsilon S_k + bk S_\varepsilon), \tag{4.16}$$

$$D_t \theta = \frac{1}{\varepsilon} S_k - \frac{k}{\varepsilon^2} S_\varepsilon. \tag{4.17}$$

In [24], we show that this system is stable and positive when a and b are such that $D_t \varphi \leq 0$. This is the case when

$$a = \frac{c_1}{c_\mu}, \quad b = 1. \tag{4.18}$$

From a numerical point of view, we can use these intermediate variables for the treatment of the source terms. More precisely, after the advection and diffusion treated as above, we introduce (ϕ, θ) and make the integration in these variables and then return to the physical variables:

$$k = \varphi^{1/(a+b)} \theta^{b/(a+b)}, \quad \varepsilon = \varphi^{1/(a+b)} \theta^{-a/(a+b)}. \tag{4.19}$$

However, experience shows that numerical integration with explicit schemes using stability condition (4.25) avoids negative values for k and ε. This means that the stability condition for Navier-Stokes is stronger than for the high-Reynolds $k - \varepsilon$ turbulence model. This is another interesting feature of wall laws. Indeed, this is not the case with low-Reynolds versions of the model.

4.2.6 Time integration

The previous spatial discretization (4.5) has been presented with a first-order time scheme but as we are targeting unsteady computations, it is important to have a precise time integration scheme. To this end, a low-storage four-step Runge-Kutta scheme is used. Let us rewrite (4.1) as

$$\frac{\partial W}{\partial t} = \text{RHS}(W), \tag{4.20}$$

where RHS contains the nonlinearities. The scheme is as follows:

$$W^0 = W^n \tag{4.21}$$

$$W^k = W^0 + \alpha_k \delta t \, \text{RHS}(W^{k-1}) \quad \text{for} \quad k = 1,..,4 \qquad (4.22)$$

$$W^{n+1} = W^4 \qquad (4.23)$$

with
$$\alpha_1 = 0.11, \, \alpha_2 = 0.2766, \, \alpha_3 = 0.5, \, \alpha_4 = 1.0, \qquad (4.24)$$

for α_k. Naturally a CFL condition is necessary for stability. More details can be found in [25].

4.2.7 Local time stepping procedure

When stationary solutions are targeted, the time integration is no longer of interest. Furthermore, to accelerate the computation a local time step for each node of the mesh is used:

$$\delta t(q^i) = \min\left(\frac{\delta x^i}{|u|+c}, \frac{\rho Pr \delta x^2}{2(\mu + \mu_t)}\right), \qquad (4.25)$$

where δx^i is the minimum height of the triangles (tetrahedra) sharing the node q^i. This formula is a generalization of a 1D stability analysis. When the time integration history is important, a global time stepping procedure is used. In this case the global time step is the minimum of the local time steps.

Wall laws and time integration As we said already, wall laws are interesting as they remove from the computational domain regions where the gradients are very large and therefore permit coarser meshes and larger time steps. Low-Reynolds number corrections may appear physically more meaningful but they lead to many numerical problems. .

For instance, with the Delery [9] compression ramp (shown below) at Mach 5 and Reynolds 10^7, meshes compatible with the local (low) Reynolds number near the wall need a first node placed at 10^{-7}m from the wall in the normal direction and between 10^{-3}m and 10^{-2}m in the tangential direction along the wall. This means an aspect ratio of 10^5. Using wall laws, we performed the same simulation with the first node at 10^{-4} m. If we use the stability criteria (4.25) presented above for local time stepping, at $CFL = 1$, this implies at least three orders of magnitude in the time step size because it corresponds to $CFL = 10^3$ for the model with low-Reynolds number corrections. In addition, the linear systems obtained on such meshes are highly ill-conditioned. Finally the generation of unstructured meshes with such a high aspect ratio is really difficult in 3D, not to speak of the problem of adaptivity.

4.2.8 Implementation of the boundary conditions

The boundary conditions presented above are implemented in weak form by replacing the given values in the boundary integrals of the weak formulation of the numerical scheme. Denotes by $(\vec{t}, \vec{n})$ the local orthonormal basis at a boundary node. The weak formulation of (4.1) contains the following integrals on boundaries:

$$\int_\Gamma W(\vec{u}.\vec{n})d\sigma, \quad \int_\Gamma p\vec{n}d\sigma, \quad \int_\Gamma p(\vec{u}.\vec{n})d\sigma, \quad \int_\Gamma (\mathbf{S}.\vec{n})d\sigma,$$
$$\int_\Gamma (\vec{u}\mathbf{S})\vec{n}d\sigma, \quad \int_\Gamma (\chi+\chi_t)\frac{\partial T}{\partial n}d\sigma,$$
$$\int_\Gamma (\mu+\mu_t)\frac{\partial k}{\partial n}, d\sigma, \quad \int_\Gamma (\mu+c_\varepsilon\mu_t)\frac{\partial \varepsilon}{\partial n}. \qquad (4.26)$$

Inflow and outflow To introduce characteristic type boundary conditions, along these boundaries the fluxes are decomposed into positive and negative parts [28] according to the sign of the eigenvalues of the Jacobian $\frac{\partial F}{\partial W}$ of the convective operator F:

$$\int_{\Gamma_\infty} F.\vec{n}d\sigma = \int_{\Gamma_\infty} (A^+ W_{in} + A^- W_\infty).\vec{n}d\sigma, \qquad (4.27)$$

where W_{in} is the computed (internal) value at the previous iteration and W_∞ the external value, given by the flow conditions. Moreover, viscous effects are neglected on these boundaries.

Symmetry and solid walls in inviscid flows Along these boundaries, only the pressure integral (second integral in (4.26)) is taken into account in the momentum equation. For Euler simulations (inviscid), a solid wall is also treated like a symmetry boundary condition.

Homogeneous Neumann boundary conditions for k and ε are natural symmetry conditions; they are easy to implement in a FVM-FEM formulation as vanishing for symmetry and in-outflow boundaries.

4.2.9 Solid walls: transpiration boundary condition

Moving walls can be replaced by an equivalent transpiration boundary condition on a fixed mean boundary. We shall use it at times to simulate shape deformations. For instance if a function ϕ is to be zero on a wall and the wall position has moved by a distance α then, by a Taylor expansion on can see that the boundary condition can still be applied at the old wall position.

But, now it has a Fourier form:

$$\phi + \alpha \frac{\partial \phi}{\partial n} = 0. \qquad (4.28)$$

In (4.26) the injection velocity is prescribed. In that case, only the implementation of the tangential velocity needed by the wall laws is modified.

4.2.10 Solid walls: implementation of wall laws

The aim here is to provide expressions for the momentum and energy equations for adiabatic and isothermal walls. To express the fourth and fifth integrals in (4.26), we split $\mathbf{S}.\vec{n}$ over $(\vec{t}, \vec{n})$.

This gives:
$$\mathbf{S}.\vec{n} = (\mathbf{S}.\vec{n}.\vec{n})\vec{n} + (\mathbf{S}.\vec{n}.\vec{t}).\vec{t}. \qquad (4.29)$$

In our implementation, the first term (S_{nn}) in the right-hand side of (4.29) has been neglected and the following wall laws have been used:

$$\vec{u}.\vec{n} = 0, \qquad (4.30)$$

$$(\mathbf{S}\vec{n}.\vec{t})\vec{t} = -\rho_w u_\tau^2 \vec{t}. \qquad (4.31)$$

The conditions above and the weak form of the energy equation give:

$$\int_\Gamma \left((\vec{u}\mathbf{S})\vec{n} + (\chi + \chi_t)\frac{\partial T}{\partial n} \right) d\sigma = 0.$$

However, in the isothermal case, the weak form of the energy equation is closed using a Reynolds hypothesis for the thermal effects:

$$\int_\Gamma \left((\vec{u}\mathbf{S})\vec{n} + (\chi + \chi_t)\frac{\partial T}{\partial n} \right) d\sigma = \int_{\Gamma_w} -1.24 \rho_w u_\tau^2 \gamma u \, d\sigma.$$

4.3 Incompressible flows

The flow solver used in this book uses the primitive variable (u, p). All the applications we consider require turbulence modelling. The turbulence modelling and wall laws are similar to what we said in the previous section for compressible flows. We will therefore give only a brief description. More details can be found in [23].

Consider the following set of incompressible Navier-Stokes and $k - \varepsilon$ equations:
Find $W = (p, u, k, \varepsilon)$ in a domain Ω, and for all times $t \in (0, T)$, solution of:

$$\begin{cases} \nabla.u = 0, \\ \frac{\partial u}{\partial t} + u\nabla u + \nabla p - \nabla \cdot S = 0, \\ \frac{\partial k}{\partial t} + u\nabla k - \nabla.((\nu + \nu_t)\nabla k) = S_k, \\ \frac{\partial \varepsilon}{\partial t} + u\nabla \varepsilon - \nabla.((\nu + c_\varepsilon \nu_t)\nabla \varepsilon) = S_\varepsilon. \end{cases} \qquad (4.32)$$

Here p and ν are the kinetic pressure and viscosity (i.e. divided by the density); S_k, S_ε and μ_t come from the corresponding expression in the compressible case presented in Chapter 2 after division by the density and removal of the terms vanishing with $\nabla \cdot u$:

$$S_k = S : \nabla u^T - \varepsilon, \qquad S_\varepsilon = C_1 S : \nabla u - C_2 \frac{\varepsilon^2}{k}. \qquad (4.33)$$

The Newtonian stress tensor is given by $S = (\nu + \nu_t)(\nabla u + \nabla u^T)$ and the various constants are given in Chapter 2.

For the applications presented in this book, these equations are solved with compatible combinations of the several boundary conditions.

- Inflow boundary condition:
$$u = u_\infty, k = k_\infty, \varepsilon = \varepsilon_\infty. \tag{4.34}$$
- Outflow boundary condition:
$$S.n.n = p, \quad S.n.t = 0, \quad \frac{\partial k}{\partial n} = \frac{\partial \varepsilon}{\partial n} = 0. \tag{4.35}$$
- Periodic boundary condition applied to all variables except the pressure.
- Boundary conditions on solid walls using wall laws.
- Boundary conditions along symmetry lines.

As for compressible flows the initial conditions are uniform values for the variables:
$$u = u_0, \quad k = k_0, \quad \varepsilon = \varepsilon_0, \quad \text{at } t = 0,$$
with small values for the latest two (say 10^{-6}). However, as is well known, an important difficulty in solving these equations is to provide an initial divergence-free velocity field $u_0(x)$. This is another advantage of wall laws compared to low-Reynolds models. Indeed, the initial solution with wall laws is uniform, so divergence-free.

4.3.1 Solution by a projection scheme

The following pressure Poisson equation (PPE) can be derived from (4.32):
$$\Delta p = -\nabla.(u.\nabla u) \text{ in } \Omega, \tag{4.36}$$
with
$$\partial_n p = n.(\nabla.S - \partial_t u - u.\nabla u), \text{ Inflow} \tag{4.37}$$
$$p = (S.n.n), \text{ outflow} \tag{4.38}$$
$$\partial_n p = 0 \quad \text{solid walls}. \tag{4.39}$$

The projection scheme [7, 8, 14, 15, 31, 32] is an iterative scheme which decouples the pressure and velocity components.

In the simplest projection scheme we use, at each iteration, we solve one explicit problem for the velocity and one implicit problem for the Poisson equation:

0. Given u^n with $\nabla.u^n = 0$,
1. Compute $\tilde{u}$ (explicit) by:
$$\tilde{u} = u^n + \delta t(-u^n.\nabla u^n + \nabla.S^n) \text{with b.c.above}, \tag{4.40}$$
2. Compute φ from
$$-\Delta \varphi = -\nabla.\tilde{u} \text{ in } \Omega, \tag{4.41}$$
$$\partial_n \varphi = 0 \text{ on all boundaries except outflow}, \tag{4.42}$$
$$\varphi = 0 \text{ outflow}. \tag{4.43}$$
3. Compute $u^{n+1} = \tilde{u} - \nabla \varphi$ and $p^{n+1} = \varphi/\delta t$.

4. $u^n \leftarrow u^{n+1}$ and go to step (1).

Here the time integration is a simple backward Euler scheme.

As $\partial_n \varphi = 0$ on walls, we therefore have $u^{n+1}.n = \tilde{u}.n$ and slip boundary conditions remain valid after projection.

4.3.2 Spatial discretization

We use the positive streamwise invariant (PSI) version of the fluctuation splitting scheme of [29] for the discretization of the convective term $(u^n \cdot \nabla u^n)$. This scheme is interesting as it is compact, accurate, robust and is reasonably easy to implement. It also guarantees the positivity of k and ε when applied to $\partial_t k + u \cdot \nabla k$ and $\partial_t \varepsilon + u \cdot \nabla \varepsilon$.

For all the terms involving second-order derivatives, including the pressure Poisson equation, a piecewise linear continuous finite element approximation has been used. The linear system of the PPE equation is solved by a conjugate gradient method with diagonal preconditioning.

As for the compressible case, let $\Omega_h = \cup_j T_j$ be a discretization by triangles or tetrahedra of the computational domain Ω and let $\Omega_h = \cup_i C_i$ be its partition into cells. Let V_h be the set of continuous piecewise linear functions on the triangulation. The solution $\tilde{u}$ is approximated by $\tilde{u}_h \in (V_h)^d, (d = 2$ or $3)$.

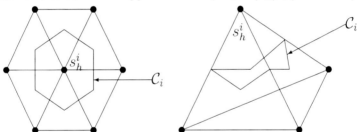

Discretization of the time derivatives All equations of the form $\partial_t v = f$ are written in weak form and discretized on the mesh with an explicit Euler scheme, giving

$$\int_{\Omega_h} \frac{1}{\delta t}(v^{n+1} - v^n)w_h = \int_{\Omega_h} f^n w_h \quad \forall w_h \in V_h. \tag{4.44}$$

With mass lumping it is

$$v_i^{n+1} = v_i^n + \frac{\delta t}{|C_i|} \int_{\Omega_h} f^n w^i, \tag{4.45}$$

where w^i is the hat function of vertex q^i. Hence at the discrete level δt is replaced by $\delta t/|C_i|$.

Discretization of right-hand sides We need to discretize in space in Ω_h:

$$u^n . \nabla u^n - \nabla . S^n, \tag{4.46}$$

with the boundary conditions given above. This is multiplied by a test function $\Phi_h \in V_h^N$ ($N = 2, 3$) and integrating by parts (4.46), we have:

$$\int_{\Omega_h} \Phi_h(u^n.\nabla u^n - \nabla.S^n)dx$$
$$= \int_{\Omega_h} u^n{}_h.\nabla u^n{}_h \Phi_h dx + \int_{\Omega_h} S^n{}_h \nabla(\Phi_h)dx - \int_{\partial\Omega_h} S^n{}_h.n\Phi_h. \quad (4.47)$$

The diffusion term is treated in the standard way, and for the convection term the PSI scheme is used. Denoting $w = u_h^n$, we have:

$$\int_{\Omega_h} w.\nabla w \Phi_h dx \sim \sum_{T \in \Omega_h} \sum_{i=1}^{d+1} \beta_i^T \sum_{l=1}^{d+1} (\frac{1}{d}\bar{w}.n_l)w_l, \quad (4.48)$$

where, in dimension d, β_i^T is the distribution coefficient for each node i of T, l denotes the local element (triangle or tetrahedra) node numbering, $\bar{w} = (\sum_{l=0}^{d+1} w_l)/(d+1)$ is the mean value of w over the element, and n_l is the inward unit normal to the triangle edge (resp. tetrahedral face) opposite to the node l multiplied by the length (resp. surface) of the corresponding triangle edge (resp. tetrahedra face).

The choice of the distribution coefficients β_i^T determines the scheme [29]. For $\beta_i^T = 1/(d+1)3$, we have a centered scheme. This approach leads therefore to an equal order discretization for velocity and pressure.

4.3.3 Local time stepping

As in the compressible case, a local time stepping procedure is introduced, based on a one-dimensional analysis of an advection diffusion equation.

$$\delta t = \frac{\delta x^2}{|u|\delta x + 2(\nu + \nu_t)}. \quad (4.49)$$

4.3.4 Numerical approximations for the $k - \varepsilon$ equations

For k and ε, we use the same algorithm as step 1 for the velocity: explicit time integration, PSI upwinding for advection using (4.48), centered P^1 finite element for diffusion and source terms:

$$k^{n+1} = k^n + \delta t(-u^n \nabla k^n + \nabla.((\nu + \nu_t^n)\nabla k^n) + S_k^n), \quad (4.50)$$

$$\varepsilon^{n+1} = \varepsilon^n + \delta t(-u^n \nabla \varepsilon^n + \nabla.((\nu + 0.7\nu_t^n)\nabla \varepsilon^n) + S_\varepsilon^n). \quad (4.51)$$

The time step comes from the Navier-Stokes stability analysis (4.49). The same approximation is used for the convection terms:

$$\int_{\Omega_h} u^n \nabla k^n \sim \sum_{T, T \in \Omega_h} \sum_{i=1}^{d+1} \beta_i^T \sum_{l=1}^{d+1} \left(\frac{1}{d}\bar{w} \cdot n_l\right) k_l^n. \quad (4.52)$$

where $\bar{w}$ again denotes the mean velocity on the element as in (4.48). A similar expression is used for ε.

Wall laws implementation

As in the compressible solver described above, integrating by part the diffusion term in the Navier-Stokes equations, the following boundary integral appears, which can be treated as in the compressible case:

$$\int_{wall} S \cdot n \Phi_h d\sigma \sim \int_{wall} u_\tau^2 t \Phi_h d\sigma, \qquad (4.53)$$

where u_τ comes from the solution of the Reichardt equation with correction for advection and pressure gradient as presented in Chapter 2.

The only point which is missing is the enforcement of the slip boundary condition $(u.n = 0)$. This is done simply by prescribing at each iteration

$$u = \text{sign}(u.t)|u|t. \qquad (4.54)$$

4.4 Mesh adaptation

We describe here an error control method through mesh adaptation. Mesh independency of results is central in simulations and obviously in design as well. Local mesh adaptation by metric control is a powerful tool for getting mesh independent results at a reduced cost [3–5, 17].

When mesh adaptation is used, not only the mesh in the computational domain, but also the number and positions of the discretization points over the shape will change during optimization. This means that, as we will see in Chapter 6, the shape and unstructured mesh deformation tools should take into account these transformations. In addition, if the shape parameterization is linked to the shape discretization, this implies a change in the design space after adaptation which we should avoid so as to keep solving the same optimization problem.

Another remark concerns the change in the mesh connectivity in Delaunay type meshes. This is difficult to take into account in the gradients because an edge swap results in a nondifferentiable change of the cost function (see chapter 5).

4.4.1 *Delaunay mesh generator*

Given a positive definite matrix $M(x)$ we define a variable metric by $\|x - y\|^2 = (x - y)^T M(x)(x - y)$. M-circles, i.e. circles with respect to the variable metric, are ellipses in the Euclidean space.

A given triangulation is said to satisfy the M-Delaunay criteria if for all inner edges the quadrangle made by its two adjacent triangles are such that the fourth vertex is outside the M-circle passing through the other three vertices.

The adaptive Delaunay mesh generator of [12] is based on these steps:

1. Discretize the boundary of Ω, the domain for which we seek a triangulation.
2. Build a M-Delaunay triangulation of the convex hull of all the boundary nodes (no internal point).

3. Add an internal point to all edges which are longer (in the variable metric) than the prescribed length.
4. Rebuild the triangulation with internal nodes, now using the Delaunay criteria and the variable metric.
5. Remove the triangles which are outside Ω.
6. Goto step 3 until the elements have the required quality.

It can be shown that the Delaunay mesh is the nearest to a quasi-equilateral mesh, in the sense that the smallest angle in the triangulation is maximized by the edge swaps so as to satisfy the Delaunay criteria. Note that if the local metric is Euclidean, the mesh elements are isotropic. So anisotropy is controlled by M, the local metric [11, 12].

4.4.2 Metric definition

As stated earlier, if we want the mesh to be adapted to the solution, we need to define the metric at each point of the domain and use it in the Delaunay algorithm above. This is usually what is required by an unstructured mesh generator having adaptation capacities [12].

The definition of the metric is based on the Hessian of the state variables of the problem. Indeed, for a P^1 Lagrange discretization of a variable u, the interpolation error is bounded by:

$$\mathcal{E} = |u - \Pi_h u|_0 \leq ch^2 |D^2 u|_0, \qquad (4.55)$$

where h is the element size, $\Pi_h u$ the P^1 interpolation of u and $D^2 u$ its Hessian matrix. This matrix is symmetric:

$$D^2 u = \begin{pmatrix} \partial^2 u/\partial x^2 & \partial^2 u/\partial x \partial y \\ \partial^2 u/\partial x \partial y & \partial^2 u/\partial y^2 \end{pmatrix} \qquad (4.56)$$

$$= \mathcal{R} \begin{pmatrix} \lambda_1 & 0 \\ 0 & \lambda_2 \end{pmatrix} \mathcal{R}^{-1}, \qquad (4.57)$$

where $\mathcal{R}$ is the eigenvectors matrix of $D^2 u$ and λ_i its eigenvalues (always real). Using this information, we introduce the following metric tensor $\mathcal{M}$:

$$\mathcal{M} = \mathcal{R} \begin{pmatrix} \tilde{\lambda}_1 & 0 \\ 0 & \tilde{\lambda}_2 \end{pmatrix} \mathcal{R}^{-1}, \qquad (4.58)$$

where

$$\tilde{\lambda}_i = \min\left(\max\left(|\lambda_i|, \frac{1}{h_{\max}^2}\right), \frac{1}{h_{\min}^2}\right), \qquad (4.59)$$

with $h_{\min}$ and $h_{\max}$ being the minimal and maximal edge lengths allowed in the mesh.

Now, if we generate, by a Delaunay procedure, an equilateral mesh with edges of length 1 in the metric $\mathcal{M}/(c\mathcal{E})$, the interpolation error $\mathcal{E}$ is equidistributed over the edges of length a_i if

$$\frac{1}{c\mathcal{E}} a_i^T M a_i = 1. \tag{4.60}$$

Three key points The previous definition is not sufficient for a suitable metric definition in the following configurations: (1) systems, (2) boundary layers, (3) multiple-scale phenomena.

Systems For systems, the previous approach leads to a metric for each variable and we should take the intersection of all these metrics. More precisely, for two metrics, we find an approximation of their intersection by the following procedure:

Let λ_i^j and v_i^j, $i, j = 1, 2$ be the eigenvalues and eigenvectors of $\mathcal{M}_j$, $j = 1, 2$. The intersection metric ($\hat{\mathcal{M}}$) is defined by

$$\hat{\mathcal{M}} = \frac{\hat{\mathcal{M}}_1 + \hat{\mathcal{M}}_2}{2}, \tag{4.61}$$

where $\hat{\mathcal{M}}_1$ (resp. $\hat{\mathcal{M}}_2$) has the same eigenvectors as $\mathcal{M}_1$ (resp. $\mathcal{M}_2$) but with eigenvalues defined by:

$$\tilde{\lambda}_i^1 = \max(\lambda_i^1, {v_i^1}^T \mathcal{M}_2 v_i^1), \quad i = 1, 2. \tag{4.62}$$

The previous algorithm is easy to extend to the case of several variables. Here, one difficulty comes from the fact that we work with variables with different physical meaning and scale (for instance pressure, density, and velocity). We will see that a relative rather than a global error estimation avoids this problem.

Boundary layers The computation of the Hessian is done by interpolation via the Green formula with Neumann boundary conditions. For instance, given ρ, ρ_x is found in W_h, a finite-dimensional approximation space, by solving approximately

$$\int_\Omega \rho_x w = -\int_\Omega \rho w_x \quad \forall w \in W_h, \tag{4.63}$$

and then similarly for ρ_{xx}.

However, this does not lead to a suitable mesh for boundary layers. Indeed, the distance of the first layer of nodes to the wall will be quite irregular. Another important ingredient, therefore, is a mixed Dirichlet-Neumann boundary condition for the different components of the metric on wall nodes for viscous computations. More precisely, the eigenvectors for these nodes are the normal and tangent unit vectors and the eigenvalue corresponding to the normal eigenvector is a prescribed value depending on the Reynolds number. The tangential eigenvalue comes from the metric of the solution.

More precisely, along the wall the previous metric $\mathcal{M}(x)$ is replaced by a new metric $\hat{\mathcal{M}}(x)$:
$$\hat{\mathcal{M}}(x) = T\Lambda T^{-1}, \qquad (4.64)$$
where
$$\Lambda = \mathrm{diag}\left(\frac{1}{h_n^2}, \lambda_\tau\right) \quad \text{and} \quad T = (\vec{n}(x), \vec{\tau}(x)). \qquad (4.65)$$

The refinement along the wall can now be monitored through h_n. This allows, for instance, for shocks and boundary layers to interact. This metric is propagated through the flow by a smoothing operator.

Multi-scale phenomena Difficulties appear when we try to compute multi-scale phenomena, such as turbulent flows, by this approach. For instance, when we have several eddies with variable energy, it is difficult to capture the weaker ones, especially if there are shocks involved. We notice that (4.55) leads to a global error while we would like to have a relative one. We propose the following estimation which takes into account not only the dimension of the variables but also their magnitude:
$$\tilde{\mathcal{E}} = \left|\frac{u - \Pi_h u}{\max(|\Pi_h u|, \epsilon)}\right|_0 \le ch^2 \left|\frac{D^2 u}{\max(|\Pi_h u|, \epsilon)}\right|_0, \qquad (4.66)$$
where we have introduced the local value of the variable in the norm. ϵ is a cut-off to avoid numerical difficulties and also to define the difference between the orders of magnitude of the smallest and largest scales we try to capture. Indeed, when a phenomena falls below ϵ, it will not be captured. This is similar to looking for a more precise estimation in regions where the variable is small. Another important observation is that this rescaling also works well with intersected metrics coming from different quantities.

4.4.3 Mesh adaptation for unsteady flows

The adaptive algorithm above uses interpolation of solutions between meshes. This is a major source of error in unsteady configurations [18, 1, 2]. The error also exists in steady cases but does not influence the final stationary solution as temporal derivatives vanish. The following transient fixed point algorithm permits us to avoid this difficulty.

Fixed point algorithm At iteration i of the adaptation, one denotes the mesh, the corresponding solution and the metric by $\mathcal{H}_i, \mathcal{S}_i, \mathcal{M}_i$. Introduce $\Delta t = \frac{\tau}{N_{\mathrm{adap}}}$ with τ the smallest time-scale of interest (the time scale of the phenomenon one would like to capture). We would like through N_{adap} and $2N_{\mathrm{adap}}$ adaptive simulations to verify if the time-dependent result is mesh-independent. The transient adaptive loop is as follows:

> initialization: $t = 0$, $\quad \mathcal{H}_0, \mathcal{S}_0, \Delta t$ given,
> **for** $(i = 0;\ i < i_{\mathrm{max}};\ i++)$
> $\quad$ Compute the metric: $\quad (\mathcal{H}_i, \mathcal{S}_i) \to \mathcal{M}_i$,

Generate the new mesh: $(\mathcal{H}_i, \mathcal{M}_i) \to \mathcal{H}_{i+1}$,
Interpolate the solution on the new mesh: $(\mathcal{H}_i, \mathcal{S}_i, \mathcal{H}_{i+1}) \to \overline{\mathcal{S}}_i$,
Advance in time by Δt: $(\mathcal{H}_{i+1}, \overline{\mathcal{S}}_i) \to \mathcal{S}_{i+1}$,
end_for (4.67)

As mentioned, the interpolations introduces perturbations when applied to time-dependent flows but also because of the time lag between the metric and the solution which increases if the time step in the numerical scheme is much smaller than Δt. One can reduce this effect by adapting more often the mesh to the solution but then the interpolation errors become important and also the computing time. Two modifications can reduce these drawbacks [1, 2, 18].

First, replace the metric based on the final solution by the interpolation (4.68) below, taking into account the changes in the solution between two successive adaptations. This introduces a combination of intermediate metrics. More precisely, if $u^p, p = n, ..., m$, are successive state iterates between two adaptations to take place at iteration n and m, one defines:

$$\hat{\mathcal{M}} = \frac{1}{m-n} \sum_{p=n}^{m} M^p(u^p), \quad (4.68)$$

where M^p is the metric based on the state at iteration p.

Then, introduce an extra fixed point iteration with NFIX internal iterations with the aim of finding a fixed point for the ensemble (metric, mesh, solution) where the metric is defined by (4.68). The previous algorithm is modified as:

$\mathcal{H}_0, \mathcal{S}_0, \Delta t,$ NFIX given, $i = 0$, $t = 0$

while $(t < t_{\max})$, **do**

$$j = 0 \quad \hat{\mathcal{M}}_i^j = \left(\mathcal{I} \left(\frac{1}{h_{max}^2} \right) \right),$$

while $\left(j < \text{NFIX} \quad \text{or} \quad ||(\frac{\partial \mathcal{S}}{\partial t})_i^{j+1} - (\frac{\partial \mathcal{S}}{\partial t})_i^j|| > TOL \right)$, **do**

- Generate the new mesh using metric intersection:

$$(\mathcal{H}_i^j, \hat{\mathcal{M}}_i^j) \to \mathcal{H}_i^{j+1}$$

- Interpolate the previous solution and its time derivative over the new mesh:

$$\left(\mathcal{H}_i^{j+1}, \mathcal{S}_i^0, \left(\frac{\partial \mathcal{S}}{\partial t} \right)_i^0, \mathcal{H}_i^j \right) \to \left(\overline{\mathcal{S}}_i^{j+1}, \overline{\left(\frac{\partial \mathcal{S}}{\partial t} \right)_i^{j+1}} \right)$$

One insists on the fact that the interpolation is from the solution at the beginning of the internal fixed point $(\mathcal{S}_i^0)$ and not from the courant solution $(\mathcal{S}_i^j)$.

- Advance the state in time by Δt and compute its contribution to metric intersection:

$$\left(\mathcal{H}_i^{j+1}, \overline{\mathcal{S}}_i^{j+1}, \overline{\left(\frac{\partial \mathcal{S}}{\partial t}\right)_i}^{j+1}\right) \to \left(\mathcal{S}_i^{j+1}, \left(\frac{\partial \mathcal{S}}{\partial t}\right)_i^{j+1}, \mathcal{M}_i^{j+1}\right)$$

- Intersect the intermediate metrics:

$$(\hat{\mathcal{M}}_i^j, \mathcal{M}_i^{j+1}) \to \hat{\mathcal{M}}_i^{j+1}$$

$j++$ **end_while**

$$\left(\mathcal{H}_{i+1}^0, \mathcal{S}_{i+1}^0, \left(\frac{\partial \mathcal{S}}{\partial t}\right)_{i+1}^0\right) \leftarrow \left(\mathcal{H}_i^j, \mathcal{S}_i^j, \left(\frac{\partial \mathcal{S}}{\partial t}\right)_i^j\right)$$

$i++$ **end_while**

4.5 An example of adaptive unsteady flow calculation

This is for the prediction of an airfoil stall; it is interesting as it shows why it is important to use accurate time integration with small time steps together with mesh adaptation. This is the case especially when stall is due to the unsteadiness in the wake (Figs. 4.1 and 4.2). Indeed, the simulation shows that in that case the stall is clearly due to the fact that the wake becomes unsteady at some incidence. Simulations are presented for a 2D profile called ONERA-A (kindly made available to us by Onera-Cert) at incidence of 7°, 13°, and 15° using the same accuracy for time and space integrations for all three cases. Stall is expected for incidence 7°. Mesh adaptation is used for the three cases;, without it stall is not predicted. As the Mach number is quite low, in principle this case is not favorable for an explicit solver.

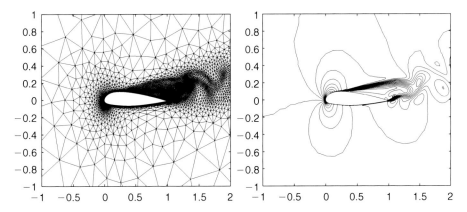

FIG. 4.1. Stall prediction for a Onera A profile. Intermediate adapted mesh and iso-Mach contours for 15° incidence.

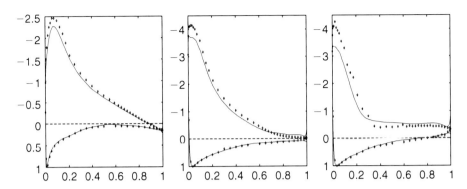

FIG. 4.2. Stall prediction for a Onera A profile. Mean pressure distribution for 7° (left), 13° (middle) and 15° (right) incidence and comparison with experiments (dots). Stall is predicted but the pressure level is too low around the leading edge.

References

[1] Alauzet, F. George, P. L. Frey, P. and Mohammadi, B. (2002). Transient fixed point based unstructured mesh adaptation, *Inter. J. Num. Methods for Fluids*, **43(6)**, 475491.

[2] Alauzet, F. Frey, P. Hecht, F. Mohammadi, B. and George, P.L. (2007). Unstructured 3D mesh adaptation for time dependent problems. Application to CFD simulations, *J. Comp. Phys.*, **222**, 269-285.

[3] Borouchaki, H. Castro-Diaz, M.J. George, P.L. Hecht, F. and Mohammadi, B. (1996). Anisotropic adaptive mesh generation in two dimensions for CFD, *Proc. 5th Int. Conf. on Numerical Grid Generation in Computational Field Simulations*, Mississipi State University.

[4] Borouchaki, H. George, P.L. and Mohammadi, B. (1996). Delaunay mesh generation governed by metric specifications. Part II: applications, *Finite Element in Analysis and Design*, special issue on mesh adaptation.

[5] Castro-Diaz, M. Hecht, F. and Mohammadi, B. (1995). Anisotropic grid adaptation for inviscid and viscous flows simulations, *Int. J. Num. Meth. Fluid*, **25**, 475-491.

[6] Coquel, F. (1994). The hybrid upwind scheme, *private communication*.

[7] Chorin, J.A. (1967). A numerical method for solving incompressible viscous flow problems, *J. Comput. Phys.* **2**, 12-26.

[8] Chorin, J.A. (1968). Numerical solution of the Navier-Stokes equations, *Math. Comput.*, **22**, 745-762.

[9] Delery, J. and Coet, M.C. (1990). Experiments on shock-wave/boundary-layer interactions produced by two-dimensional ramps and three-dimensional obstacles, *Proc. Workshop on Hypersonic Flows for Reentry Problems*, Antibes.

[10] Dervieux, A. (1985), Steady Euler simulations using unstructured meshes, *VKI lecture series*, **1884-04**.

[11] George, P.L. Hecht, F. and Saltel, E. (1990). Fully automatic mesh generator for 3d domains of any shape, *Impact Comp. Sci. Eng.* **2**, 187-218.

[12] George, P.L. (1991). *Automatic Mesh Generation. Applications to Finite Element Method*, John Wiley, London.

[13] Godlewski, E. and Raviart, P.A. (1997). *Nonlinear Hyperbolic Equations*, Ellipses, Paris.

[14] Gresho, P.M. (1990). On the theory of semi-implicit projection methods for viscouc incompressible flow and its implementation via a finite element method that also introduces a nearly consistent mass matrix I: theory, *Int. J. Num. Meth. Fluids*, **11**, 587-620.

[15] Gresho, P.M. (1991). Some current CFD issues relevant to the incompressible Navier-Stokes equations, *Comput. Meth. Appl. Mech. Eng.*, **87**, 201-252.

[16] Harten, A. Roe, P. and Van Leer, B. (1983). On upstream differencing and Godunov-type scheme for hyperbolic conservation laws, *Siam Rev.*, **25**, 35-61.

[17] Hecht, F. and Mohammadi, B. (1997). Mesh adaptation by metric control for multi-scale phenomena and turbulence, *American Institute of Aeronautics and Astronautics*, **97-0859**.

[18] Hecht, F. and Mohammadi, B. (2001). Mesh adaptation for time dependent simulation, optimization and control, *Revue Européenne des éléments finis*, **10(5)**, 575-595.

[19] Hughes, T.J.R. Franca, L.P. and Mallet, M. (1986). A new finite element formulation for computational fluid dynamics: I. symmetric forms of the compressible Euler and Navier-Stokes equations and the second law of fhermodynamics, *Comput. Meth. Appl. Mech. Eng.*, **54**, 223-234.

[20] Hughes, T.J.R. and Mallet, M. (1986). A new finite element formulation for computational fluid dynamics: III. The generalized streamline operator for multidimensional advective-diffusive systems, *Comput. Meth. Appl. Mech. Eng.*, **58**, 305-328.

[21] Hughes, T.J.R. and Mallet, M. (1986). A new finite element formulation for computational fluid dynamics: IV. Discontinuity capturing operator for multi-dimensional advective-diffusive systems, *Comput. Meth. Appl. Mech. Eng.*, **58**, 329-339.

[22] Larrouturou, B. (1989). How to preserve the mass fraction positivity when computing compressible multi-component flows, *INRIA report* **1080**.

[23] Medic, G. and Mohammadi, B. (1998). NSIKE unstructured solver for laminar and turbulent incompressible flows simulation, *INRIA report* **3644**.

[24] Mohammadi, B. and Pironneau, O. (1994). *Analysis of the k-epsilon Turbulence Model*, John Wiley, London.

[25] Mohammadi, B. (1994). CFD with NSC2KE : a user guide, *Technical report INRIA* **164**.

[26] Pironneau, O. (1990),*Finite Element in Fluids*, John Wiley, London.

[27] Roe, P.L. (1981). Approximate Riemann solvers, parameters vectors and difference schemes, *J.Comp. Phys.* **43**, 357372.
[28] Steger, J. and Warming, R.F. (1983). Flux vector splitting for the inviscid gas dynamic with applications to finite-difference methods, *J. Comp. Phys.* **40**, 263-293.
[29] Struijs, R. Deconinck, H. de Palma, P. Roe, P. and Powel, G.G. (1991), Progree on multidimensional upwind Euler solvers for unstructured grids, *American Institute of Aeronautics and Astronautics*, **91-1550**.
[30] Van Albada, G.D. and Van Leer, B. (1984). Flux vector splitting and Runge Kutta methods for the Euler equations, *ICASE* **84-27**.
[31] Temam, R. (1984), Navier-Stokes Equations, North-Holland - Elsevier, Amsterdam.
[32] Tuomela, J. and Mohammadi, B. (2005). Simplifying numerical solution of constrained PDE systems through involutive completion, *M2AN*, **39(5)**, 909929.
[33] Yee, H.C. (1987). Upwind and symmetric shock-capturing schemes, *NASA TM* **89464**.

5

SENSITIVITY EVALUATION AND AUTOMATIC DIFFERENTIATION

5.1 Introduction

This chapter is devoted to the presentation of various approaches for the evaluation of the gradient of functionals. We also discuss automatic differentiation (AD) of computer programs as a tool for shape optimization.

For gradient-based shape optimization methods, it is necessary to have an estimation of the derivatives of the discrete cost function with respect to control parameters. It is clear that when the number of control parameters is large, an adjoint equation is necessary [41,28,7,20]. It is tempting to use a discretization of the adjoint equation of the continuous problem; this however would not account for the discretization errors of the numerical schemes (like numerical dissipation for instance). Automatic differentiation produces the exact derivatives of the discrete cost function. Moreover, in reverse mode, the cost of this evaluation is independent of the number of control parameters as for a standard adjoint method.

AD can also be used to analyze the sensitivity of the numerical scheme itself [26] to various parameters such as the different numerical fluxes for finite volume methods for compressible inviscid flows. We used Roe and Osher fluxes with and without MUSCL second-order reconstruction [42, 13, 48] for the solution of the Euler equations. We showed that this approach works equally well on parabolic and hyperbolic equations. Viscous turbulent configurations have also been investigated [35]: a $k-\varepsilon$ model [29] with special wall laws including pressure and convection effects in the Jacobian.

Therefore, AD is not only a tool for the evaluation of sensitivities, but it also helps to understand the state equation of a problem and the contribution of each operator involved in the computation of the sensitivities, and this at a reasonable cost.

In multicriteria optimization sensitivity analysis is important to discriminate between Pareto points and this even if a gradient free approach is used. Indeed, knowledge of sensitivity permits us to qualify various points of a Pareto front from the point of view of robustness: two points on a Pareto front can be compared if one considers the sensitivity of the functional with respect to the independent variables which are not control parameters. The robust optimum is the one with lowest sensitivity.

Also, sensitivity evaluation is important because often in simulations information on the uncertainties on the results is more important than the results themselves. For instance, it is essential to be able to identify dominant independent variables in a system, as these will need more accurate monitoring and for which precise measurements should be provided.

These concepts are central in Validation and Verification (V&V) issues which refer to all of the activities that are aimed at making sure that the software will function as required. Indeed, it is important to include robustness issues into the specifications using sensitivity analysis and see that a simulation should therefore be seen as multicriteria minimization.

Consider the following simulation loop linking a set of independent to dependent variables and eventually leading to the calculation of a functional $J(x, \varphi, h)$, a function of independent variables x_i, for instance, a geometric parametrization, φ the physical variables defining the flow and h the solution procedure parameters (discretization, accuracy, convergence rates, etc.)

$$(x, \varphi, h) \to q(x, h) \to U(\varphi, q(x), h) \to J(x, \varphi, h, q(x, h), U(\varphi, q(x, h))). \quad (5.1)$$

Flow calculations enter this class with, in x the parameters defining the geometry (e.g. the cord of an airfoil), q for the geometric quantities (mesh, vertices, etc.) and U field flow variables solution of the state equation $F(U(q(x, h), \varphi, h) = 0$. For a geometrical set of parameters (x^*) and flow conditions (φ^*), the solution of the state equation can be seen as minimization of

$$J_1(x^*, \varphi^*, h) = \|F(U(q(x^*, h), \varphi^*, h)\| \quad (5.2)$$

Hence, we look for the best solution procedure which minimizes the residual: $h = \mathrm{argmin}_{h \in H} J_1(x^*, \varphi^*, h)$. The admissible space H includes the constraint on the solution procedure (e.g. maximum number of vertices one can afford, accuracy in the solution of nonlinear and linear systems,...).

Robustness issues can now be introduced through control of the sensitivity with respect to the other independent variables around the functioning point (x^*, φ^*):

$$J_2(x^*, \varphi^*, h) = \|\nabla_{x, \varphi} J_1(x^*, \varphi^*, h)\|. \quad (5.3)$$

The simplest way to proceed is by penalizing these sensitivities and looking for the solution of $h = \mathrm{argmin}_{h \in H} J_1(x^*, \varphi^*, h) + J_2(x^*, \varphi^*, h)$. This is because no calculation can be reliable if it is too sensitive to small perturbations in the data. This also shows that the solution procedure probably needs changes once it includs robustness issues, except if the optimum of the constrained and unconstrained problems are the same, which is quite unlikely. It is obvious that the cost of this approach makes it difficult to use in practice; one should however pay attention and at least a posteriori evaluate the sensitivity of the solution to perturbation of the independent variables [47]. We will also discuss this issue when presenting global optimization algorithms in Chapter 7.

Another interest in sensitivity evaluation is to build response surfaces. These can be built by various approaches (see Chapter 7) such as radial basis functions or Kriging or even using local Taylor or Padé expansions. In all cases, information on sensitivities improves the accuracy of the construction. Once these reduced order models are built, they can also be used, as mentioned above, for robustness analysis through Monte Carlo simulations. One needs then to make hypotheses on the probability function distributions of the independent variables which are not control parameters. Finally, if response surfaces are not available, knowing the gradients, Monte Carlo simulations can be avoided using statistical approaches such as first or second-order reliability methods (FORM, SORM) [30, 31] giving an indication of the probability of failure of the design, or the method of moments [27] which produces information on the mean, variance, etc. for the functional, for instance.

5.2 Computations of derivatives

Let us recall four different approaches to finding the derivatives of a cost function J with respect to control parameters $x \in R^N$, in the case when $J(x, U(x))$ depends on x also via a function $U(x)$ which is the solution of a PDE, the state equation:

$$J(x, U(x)) \text{ where } x \to U(x) \text{ is defined by } E(x, U(x)) = 0. \qquad (5.4)$$

5.2.1 Finite differences

The easiest approach is of course finite differences. Indeed, we need here only multiple cost function evaluations and no additional coding. It is well suited to black-box code users who do not have access to the source code:

$$\frac{dJ}{dx_i} \approx \frac{1}{\epsilon}[J(\vec{x} + \epsilon\vec{e_i}, U(\vec{x} + \epsilon\vec{e_i})) - J(\vec{x}, U(\vec{x}))]. \qquad (5.5)$$

However, there are three well-known difficulties :

- the choice of ϵ (especially for multicriteria optimizations);
- the round-off error due to the subtraction of nearly equal terms;
- a computing cost proportional to the size of the state space times the control space.

5.2.2 Complex variables method

To avoid a critical choice for ϵ and also to avoid the subtraction error present in finite differences, we can use the complex variables method [45, 1]. Indeed as J is real valued we have

$$J(x_i + \mathrm{i}\epsilon, U(x_i + \mathrm{i}\epsilon)) = J(x, U(x)) + \mathrm{i}\epsilon J'_{x_i} - \frac{\epsilon^2}{2} J''_{x_i x_i} - \mathrm{i}\frac{\epsilon^3}{6} J'''_{x_i x_i x_i} + o(\epsilon^3),$$

$$\text{implies} \quad J_{x_i} = \frac{\mathrm{Im}(J(x_i + \mathrm{i}\epsilon, U(x_i + \mathrm{i}\epsilon)))}{\epsilon} + o(\epsilon) \qquad (5.6)$$

where $x_i + \mathrm{i}\epsilon$ means add to the ith component of the control parameters the increment $\epsilon\sqrt{-1}$. We can see that there is no more subtraction and the choice of ϵ is less critical.

In the same way, the second derivative can be found as:

$$\frac{d^2 J}{dx^2} = -2 \frac{Re(J(x_i + i\epsilon, U(x_i + i\epsilon))) - J(x, U(x))}{\epsilon^2}. \quad (5.7)$$

Unfortunately, here, there is again a subtraction, but less important than when using a central difference formula ($J''_x = ((J(x+\epsilon) - 2J(x) + J(x-\epsilon))/\epsilon^2)$).

In practice, this method requires a redefinition in the computer program of all variables and functions from *real* to *complex*. The computational complexity of the approach is comparable to second-order finite differences, since complex operations and storage require at least twice the effort compared with reals. Here, too, the complexity is also proportional to the number of states times control variables.

5.2.3 State equation linearization

To reduce the influence of ϵ one way is to use calculus of variation and compute with finite differences only the partial derivatives of functions. Thus, denote $\delta x = \epsilon \vec{e^i}$ and let δU be the variation of U (i.e. $\delta U = \delta x_i \partial_{x_i} U$). By linearization of $E(x, U(x)) = 0$:

$$\frac{\partial E}{\partial U} \delta U = -\frac{\partial E}{\partial x} \delta x \approx -[E(x + \epsilon \vec{e_i}, U(x)) - E(x, U(x))]. \quad (5.8)$$

For fluids most numerical implicit solvers are based on a semi-linearization of the equations as for instance in the following quasi-Newton method using Krylov subspaces:

$$\frac{\partial E^n}{\partial U}(U^{n+1} - U^n) = -E(x, U^n(x)). \quad (5.9)$$

In other words, to solve (5.8), we can use the same implicit solver but with a different right-hand side. More precisely, we have to solve (5.9) N times the dimension of x, and each solution gives one column of $\partial_x U$. After substitution, we have:

$$\frac{dJ}{dx} = \frac{\partial J}{\partial x} + \frac{\partial J}{\partial U} \frac{\partial U}{\partial x}, \quad (5.10)$$

where the partial derivatives are computed by finite differences but $\partial U/\partial x$ is computed by solving (5.8). But the computing cost problem remains proportional to N times the dimension of U.

5.2.4 Adjoint method

In the former method we have:

$$\frac{dJ}{dx} = \frac{\partial J}{\partial x} - \frac{\partial J}{\partial U}\left(\left(\frac{\partial E}{\partial U}\right)^{-1} \frac{\partial E}{\partial x}\right) = \frac{\partial J}{\partial x} - \left(\left(\frac{\partial E}{\partial U}\right)^{-T} \frac{\partial J}{\partial U}\right) \frac{\partial E}{\partial x}. \quad (5.11)$$

In this expression, we only gathered in a different way the three terms involved in the derivative of the cost function. This simple action is essential, as we can now introduce a new variable p, called the adjoint variable, such that

$$\left(\frac{\partial E}{\partial U}\right)^T p = \frac{\partial J}{\partial U}, \tag{5.12}$$

which makes it possible to compute the gradient at a cost independent of N, as only one solution of (5.12) is required:

$$\frac{dJ}{dx} = \frac{\partial J}{\partial x} - p^T \frac{\partial E}{\partial x}. \tag{5.13}$$

This can also be linked with a saddle-point problem for the Lagrangian.

5.2.5 Adjoint method and Lagrange multipliers

Introduce the Lagrangian $L = J + p^T E$, and write stationarity with respect to U, p, x. This gives an equation for p:

$$\frac{\partial L}{\partial U} = \frac{\partial J}{\partial U} + p^T \frac{\partial E}{\partial U} = 0, \tag{5.14}$$

the state equation, and

$$\frac{dJ}{dx} = \frac{\partial L}{\partial x} = \frac{\partial J}{\partial x} + p^T \frac{\partial E}{\partial x}. \tag{5.15}$$

Notice that (5.9) is almost similar to (5.14) and differs only by a transposition of the Jacobian matrix.

Here, the cost is proportional to the state space size plus (and not times) the control space size, because the state and adjoint equations are solved once only; but we still need to compute $\partial E/\partial x$. This can be easily done using finite differences or by AD in direct mode (see below) [37].

This adjoint method can be applied at the level of the continuous problem or at the level of the discrete problem [41, 28, 34, 39, 11, 12]. The continuous level has been widely used for various optimization and shape design problems. If we choose to linearize the state equation at the continuous level, two difficulties arise:

- it does not take into account numerical errors of the discretization;
- when the equations are complicated (for instance flow system with turbulence modeling coupled to an elasticity system), it becomes difficult to do the derivation also at the continuous level and produce bug-free code.

The idea is therefore to work at the discrete level and in an automatic fashion.

5.2.6 Automatic differentiation

The direct mode The leading principle is that a function given by its computer program can be differentiated by differentiating each line of the program. So the various methods of AD differ mostly in the way this idea is implemented. A review article on these can be found in [18]. The idea being simple it is best understood by means of a simple example.

Consider the function J given below and the problem of finding its derivative with respect to u at $u = 2.3$:

$$J'(u) \quad \text{where} \quad \begin{aligned} x &= 2u(u+1) \\ y &= x + \sin(u) \\ J &= x * y. \end{aligned}$$

Above each line of code insert the differentiated line:

$$dx = 2 * u * du + 2 * du * (u+1)$$
$$\mathbf{x = 2 * u * (u+1)}$$
$$dy = dx + \cos(u) * du$$
$$\mathbf{y = x + \sin(u)}$$
$$dJ = dx * y + x * dy$$
$$\mathbf{J = x * y.}$$

The derivative will be obtained by running this program with $u = 2.3, du = 1$, and $dx = dy = 0$ at the start time.

Loops and branching statements are no more complicated. Indeed, an `if` statement

```
A; if ( bool ) then B else C; D;
```

is in fact two programs. The first one: A; B; D; gives A';A;B';B; D';D; and the second one A; C; D; gives A';A; C';C; D';D; they can be recombined into

```
A';A; if ( bool) then { B';B} else {C';C} end if; D';D
```

Similarly, a loop statement like:

```
A; for i:=1 to 3 do B(i); D;
```

is in fact:

```
A; B(1);B(2);B(3); D;
```

which gives rise to:

```
A';A; B'(1);B(1);...;B'(3);B(3); D';D;
```

which is also recombined into

`A';A; for i:=1 to 3 do{ B'(i);B(i);} D';D`.

However, if the variable "bool" and/or if the bounds of the `for` statement depend on u then there is no way to take that into account. It must also be said that the function is nondifferentiable with respect to these variables.
There are also some functions which are nondifferentiable, such as $\sqrt{x}$ at $x = 0$. Any attempt to differentiate them at these values will cause an overflow or a NaN (not a number) error message.

When there are several parameters the method remains essentially the same. For instance consider $(u_1, u_2) \to J(u_1, u_2)$ defined by the program

$$y_1 = l_1(u_1, u_2) \quad y_2 = l_2(u_1, u_2, y_1) \quad J = l_3(u_1, u_2, y_1, y_2) \quad (5.16)$$

Apply the same recipe

$$dy_1 = \partial_{u_1} l_1(u_1, u_2) dx_1 + \partial_{u_2} l_1(u_1, u_2) dx_2$$
$$\mathbf{y_1} = \mathbf{l_1}(\mathbf{u_1}, \mathbf{u_2})$$
$$dy_2 = \partial_{u_1} l_2 dx_1 + \partial_{u_2} l_2 dx_2 + \partial_{y_1} l_2 dy_1$$
$$\mathbf{y_2} = \mathbf{l_2}(\mathbf{u_1}, \mathbf{u_2}, \mathbf{y_1})$$
$$dJ = \partial_{u_1} l_3 dx_1 + \partial_{u_2} l_2 dx_2 + \partial_{y_1} l_3 dy_1 + \partial_{y_2} l_3 dy_2$$
$$\mathbf{J} = \mathbf{l_3}(\mathbf{u_1}, \mathbf{u_2}, \mathbf{y_1}, \mathbf{y_2})$$

Run the program twice, first with $dx_1 = 1, dx_2 = 0$, then with $dx_1 = 0, dx_2 = 1$, or, better, duplicate the lines $dy_i =$ and evaluate both at once with $dx_1 = \delta_{ij}$, meaning that

$$d1y_1 = \partial_{u_1} l_1(u_1, u_2) d1x_1 + \partial_{u_2} l_1(u_1, u_2) d1x_2$$
$$d2y_1 = \partial_{u_1} l_1(u_1, u_2) d2x_1 + \partial_{u_2} l_1(u_1, u_2) d2x_2$$
$$\mathbf{y_1} = \mathbf{l_1}(\mathbf{u_1}, \mathbf{u_2})$$
$$d1y_2 = \partial_{u_1} l_2 d1x_1 + \partial_{u_2} l_2 d1x_2 + \partial_{y_1} l_2 d1y_1$$
$$d2y_2 = \partial_{u_1} l_2 d2x_1 + \partial_{u_2} l_2 d2x_2 + \partial_{y_1} l_2 d2y_1$$
$$\mathbf{y_2} = \mathbf{l_2}(\mathbf{u_1}, \mathbf{u_2}, \mathbf{y_1})$$
$$d1J = \partial_{u_1} l_3 d1x_1 + \partial_{u_2} l_2 d1x_2 + \partial_{y_1} l_3 d1y_1 + \partial_{y_2} l_3 d1y_2$$
$$d2J = \partial_{u_1} l_3 d2x_1 + \partial_{u_2} l_2 d2x_2 + \partial_{y_1} l_3 d2y_1 + \partial_{y_2} l_3 d2y_2$$
$$\mathbf{J} = \mathbf{l_3}(\mathbf{u_1}, \mathbf{u_2}, \mathbf{y_1}, \mathbf{y_2})$$

is evaluated with $d1x_1 = 1, \ d1x2 = 0, \ d2x_1 = 0, \ d2x_2 = 1$.

The reverse mode The reference in terms of efficiency for the computation of partial derivatives is the so-called reverse or adjoint mode. Consider a simple model problem where J is a function of two parameters and is computed by a program that uses two intermediate variables:

$$y_1 = l_1(u_1, u_2)$$
$$y_2 = l_2(u_1, u_2, y_1)$$
$$J = l_3(u_1, u_2, y_1, y_2)$$

Let us build the Lagrangian by associating to each intermediate variable a dual or adjoint variable p, except for the last line for which $p = 1$:

$$L = p_1[y_1 - l_1(u)] + p_2[y_2 - l_2(u, y_1))] + J - l_3(u, y_1, y_2). \quad (5.17)$$

Stationarity with respect to y_2, y_1 (in that order) gives

$$0 = p_2 - \frac{\partial l_3}{\partial y_2}(u, y_1, y_2)$$
$$0 = p_1 - p_2 \frac{\partial l_2}{\partial y_1}(u, y_1) - \frac{\partial l_3}{\partial y_1}(u, y_1, y_2).$$

This gives p_2 first and then p_1 and then J'_u is

$$\frac{\partial J}{\partial u_i} = p_1 \frac{\partial l_1}{\partial u_i} + p_2 \frac{\partial l_2}{\partial u_i} + \frac{\partial l_3}{\partial u_i}. \quad (5.18)$$

The difference with the direct approach is that, whatever the number of independent variables, the adjoint variables p_i are evaluated once only and so the complexity of the computation is much less. On the other hand, the method requires a symbolic manipulation of the program itself and so it is not easy to implement as a class library [2].

5.2.7 A class library for the direct mode

There are currently several implementations of the reverse mode, `Adol-C` and `Odyssée` in particular. But these may not be so easy to use by the beginner: some learning is required.

A very simple implementation of the direct mode can be done in C++ by using operator overloading. It is certainly not a good idea to use this method extensively on problems which have more than 50 control variables or so, but it is so easy and handy that it is best to be aware of it. It is possible to do the same in `Fortran 90` as well (see Maikinen [33] for example).

Principle of programming Consider again

$$J'(u) \quad \text{where} \quad \begin{aligned} x &= 2u(u+1) \\ y &= x + \sin(u) \\ J &= x * y. \end{aligned}$$

Each differentiable variable will store two numbers: its value and the value of its derivative. So we may replace all variables by an array of size 2. Hence the program becomes

```
float J[2],y[2],x[2],u[2];
// dx = 2 u du + 2 du (u+1)
x[1] = 2 * u[0] * u[1] + 2 * u[1] * (u[0] + 1);
// x = 2 u (u+1)
x[0] = 2 * u[0] * (u[0] + 1);
// dy = dx + cos(x) dx
y[1] = x[1] + cos(u[0])*u[1];
// y = x + sin(x)
y[0] = x[0] + sin(u[0]);
J[1] = x[1] * y[0] + x[0] * y[1];
// J = x * y
J[0] = x[0] * y[0];
// dJ = y dx + x dy
J[1] = x[1]*y[0] + x[0]*y[1]
```

Now, following [25] we create a C++ class whereby each variable contains the array of size two just introduced and we redefine the standard operations of linear algebra by giving our own definition such as the one used below for the multiplication:

```
#include <iostream.h>

class dfloat{
  public: float v[2];
  dfloat(){ v[1]=0;} // intialize derivative to 0
  dfloat(double a) { v[0] = a; v[1]=0;}
// above: promote double into dfloat

  dfloat& operator=(const dfloat& a)
  { v[0] = a.v[0]; v[1] = a.v[1]; return *this;}
friend dfloat operator * (const dfloat& a,
  const dfloat& b)
  { dfloat c;
  c.v[1] = a.v[1] * b.v[0] + a.v[0] * b.v[1];
  c.v[0] = a.v[0] * b.v[0];
  return c;
  }
friend dfloat operator + (const dfloat& a,
  const dfloat& b)
  { dfloat c;
  c.v[1] = a.v[1] + b.v[1];
  c.v[0] = a.v[0] + b.v[0];
  return c;
  }
// ...
```

```
void init(float x0, float dx0){ v[0]=x0; v[1] = dx0;}
};

void main () { dfloat x,u;
  u.init(2.3,1); /* Derivative w/r to u
  at u=2.3 requested */
  x = 2. * u * (u + 1.);
//...
  cout << x.v[0] <<'\t'<< x.v[1] << endl;
}
```

This program works as it is (gives 15.18 and 11.2 as expected) but to be complete and applicable to more general programs all the operators like $-, /, >, \sin, \log$, etc. must be added. The best way is to put the class definition into a file "dfloat.h" for instance. Then any C program can be differentiated (**Fortran 77** also via **f2c**, but for **C++** programs there can be conflicts with other class structures). Now all we have to do is to change all "float" (and/or double) variables into dfloat variables and add a line like "u.init(u0,1)" above to indicate that the derivatives are taken with respect to u at $u = u0$.

Implementation as a C++ class library To handle partial derivatives, the C++ implementation is done with a template class because the number of partial derivatives required is known only at compile time:

```
#include <iostream.h>

template <int N> class dfloat{
  public: float v[N];
  dfloat()
  { for(int i=1;i<N;i++) v[i]=0;}
  dfloat(double a)
  { v[0] = a; for(int i=1;i<N;i++) v[i]=0;}
  dfloat& operator=(const dfloat& a)
  { for(int i=0;i<N;i++) v[i]=a.v[i];
  return *this;}
friend dfloat operator * (const dfloat& a,
  const dfloat& b)
  { dfloat c;
  for(int i=1;i<N;i++) c.v[i] = a.v[i] * b.v[0]
  + a.v[0] * b.v[i];
  c.v[0] = a.v[0] * b.v[0];
  return c;
  }
friend dfloat operator + (const dfloat& a,
  const dfloat& b)
  { dfloat c;
```

```
    for(int i=N-1;i>=0;i--) {
    c.v[i] = a.v[i] + b.v[i];
    }
    return c;
    }
// ...
void init(float x0, int n){ v[0]=x0; v[n] = 1;}
};

void main () { dfloat<3> x,u,v;
    u.init(2.3,1);
    v.init(0.5,2);
    x = 2. * u * ( u + v);
//...
    cout << x.v[0] <<'\t'<< x.v[1]<<'\t'<< x.v[2] << endl;
}
```

In this example the partial derivatives with respect to u and v of $2u(u+v)$ are computed at $u = 2.3, v = 0.5$.

It is better to use a template class because N, which is the number of parameters, should not be fixed at a default value otherwise the library has to be edited by hand each time it is used. Templates are efficient, but here all functions with `for` statements should be implemented outside the class definition for optimal Inlining.

As before, this program works (the answer is 12.88, 10.2, 4.6) but to make it general all the operations of algebra and all the usual functions like sin( ) should be added.

In many problems, however, N may not be known at compile time; this is the case of OSD where N is the number of discretization parameters which define the boundary; then dynamic arrays cannot be avoided.

```
class dfloat{
  public: float* v; int N;
  dfloat();
  ~dfloat();
  friend dfloat operator * (dfloat& a, dfloat& b)
  { dfloat c;
  for(int i=1;i<N;i++)
  c.v[i] = a.v[i] * b.v[0] + a.v[0] * b.v[i];
  c.v[0] = a.v[0] * b.v[0];
  return c;
  }
  ...
};
```

where the class dfloat now has a constructor and a destructor to allocate and destroy the array v[ ].
```
dfloat::dfloat(){ v = new float[N];
for(int i=1;i<N;i++)v[i]=0;} // constructor
dfloat::~dfloat{ delete [] v;} // destructor
```

The problem then is that a large number of temporary arrays are created dynamically at execution time, like c.v in the multiplication above, and that takes a lot of computing time.

An optimization can be found in [3] which uses expression templates and traits and which considerably reduces the creation of temporaries so as to arrive at performances similar to those obtained with the template class library explained above. Still, to the user the simplicity of the class library is kept and it is extremely easy to link any C program to these libraries for a sensitivity analysis with respect to a few parameters.

In [2] extensive comparisons have been made between this approach (the forward mode) and the best reverse mode programs; these indicate that this approach is hard to beat when the number of control parameters is less than 50 or so.

Source codes of these classes are available at www.ann.jussieu.fr/pironneau.

5.3 Nonlinear PDE and AD

This is an application of AD to the solution of a nonlinear PDE, here the Navier-Stokes equations in conservation form. We use AD in direct mode for the evaluation of the Jacobian of the flux and use it in a quasi-Newton algorithm [14].

To solve a nonlinear PDE with a Newton type algorithm, we need the Jacobian of the operator or its action on a vector. This can be done by linearization or exact computation in the case of differentiable operators, or simply by finite differences (FD).

We consider the following form of the Navier-Stokes equations:

$$F(W) = \frac{\partial W}{\partial t} + \nabla.(f(W)) = 0, \qquad (5.19)$$

where W is the vector of conservation variables, and f represents the advective and viscous operators. This system has four equations in 2D for five variables and the system is closed using a state law (perfect gas in our case).

W^0 being the initial state, a first-order Taylor expansion leads to:

$$F(W^1) = F(W^0) + \frac{\partial F}{\partial W}(W^0).(W^1 - W^0) + o(W^1 - W^0). \qquad (5.20)$$

By requiring $F(W^1) = 0$ and neglecting the small terms, we obtain the linearized Newton algorithm:

$$W^1 = W^0 - \left(\frac{\partial F}{\partial W}\right)^{-1} . F(W^0). \tag{5.21}$$

We use a Krylov subspace approach (without preconditioner) to solve this linear system [14] and AD is used to compute $\frac{\partial F}{\partial W}$.

For the Newton scheme to converge, we need W^0 to be not too far from the solution W^1. But, in our computations, we always start from a uniform state. Therefore, a restart process is also used.

We present results for a transonic viscous computation around a NACA0012 at Mach 0.85 and Reynolds 1000. Global time stepping has been applied to reduce the mesh effect on the condition of the matrix. The mesh has around 1000 points which makes a problem with 4000 unknowns. The code was run over 350 iterations in time.

The numerical values of the different components (four conservation variables) of the linear tangent Jacobian are shown (i.e. the Jacobian times an increment δW: $(\frac{\partial f}{\partial W})\delta W$ for 100 points in the mesh obtained using FD and AD). The increment ε used for FD computations has been chosen so that the first and second-order finite differencing give the same results:

$$\begin{aligned}\frac{dF}{dW}.\delta W &\sim \frac{F(W + \varepsilon\delta W) - F(W - \varepsilon\delta W)}{2\varepsilon}, \\ \frac{dF}{dW}.\delta W &\sim \frac{F(W + \varepsilon\delta W) - F(W)}{\varepsilon}.\end{aligned} \tag{5.22}$$

These gradients are computed from the same state with a CFL number of 100. Of course, during the computation they will diverge from each other. This validates the program produced by AD. We can see that the FD approach always produces smoother gradients (see Fig. 5.1).

In Figs. (5.3) and (5.2), we show the convergence of the quasi-Newton algorithm during the computation. The Krylov space dimension is 20. Two CFL numbers of 10^2 and 10^6 have been tried. For the former case only the use of the Jacobian obtained with the automatic differentiation leads to good convergence. The case with CFL= 10^6 corresponds to the resolution of Navier-Stokes equations after removing the temporal term. This case can be seen as a resolution with GMRES with 350 (number of time step) restarts. The convergence history for the previous case and an explicit computation (CFL = 1) are shown in Fig. 5.3.

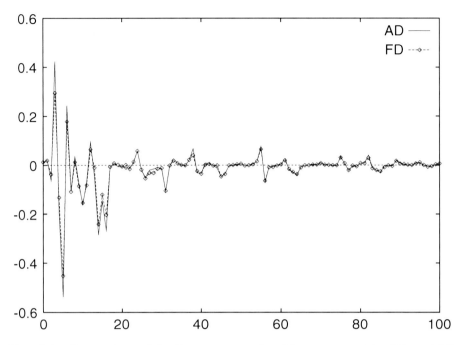

FIG. 5.1. Comparison of the linear tangent Jacobian computed by FD and AD for ρ for 100 nodes around the shape.

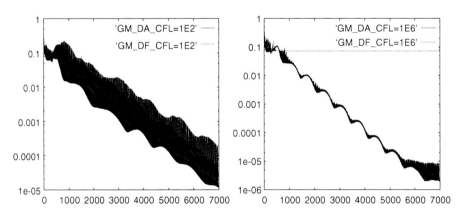

FIG. 5.2. GMRES convergence for CFL numbers of 10^2 and 10^6.

5.4 A simple inverse problem

We consider here the solution of the Burgers equation with right-hand side [10]. This control problem has been suggested to us by Prof. M. Hafez from UC Davis.

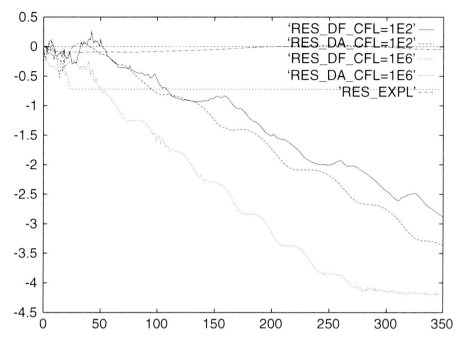

FIG. 5.3. Global convergence history. Explicit vs. implicit using FD or AD for the Jacobian evaluation. For small CFL numbers (CFL = 10^2) both the FD and AD approaches converge while for CFL = 10^6 only AD works.

$$\frac{\partial u}{\partial t} + \frac{\partial}{\partial x}\frac{u^2}{2} = 0.3xu, \qquad (5.23)$$

$$u(t,-1) = 1, u(t,1) = -0.8, \quad u(0,x) = -0.9x + 0.1. \qquad (5.24)$$

The steady solution of (5.23) is piecewise parabolic in a smooth region and has a jump (see Fig. 5.4).

$$\begin{aligned} u(x) &= 0.15x^2 + 0.85 \quad \text{for } x < x_{\text{shock}}, \\ u(x) &= 0.15x^2 - 0.95 \quad \text{for } x > x_{\text{shock}}, \end{aligned} \qquad (5.25)$$

and the shock position is found by asking for the flux to have no jump:

$$u^{-}_{\text{shock}} = -u^{+}_{\text{shock}} \quad \text{therefore} \quad x_{\text{shock}} = -\sqrt{1/3}. \qquad (5.26)$$

We use an explicit solver with Roe flux [42] for the discretization. This flux is nondifferentiable because of the presence of an absolute value. The direct solver for (5.23) and the sensitivities obtained using direct and reverse modes of AD are shown in this chapter. We also show the application of inter-procedural differentiation on this example.

This example shows that even if the target solution is simple (piecewise linear here shown in figures 5.9, 5.10 and 5.11), the optimization algorithm can fail (see Fig. 5.5); general shape optimization problems based on PDE solutions have the same characteristics.

We want to see how accurate are the derivatives produced by AD compared to finite differences. We choose (5.22) to evaluate the derivatives for FD. The cost of one evaluation is therefore 101 solutions of the Burgers equation.

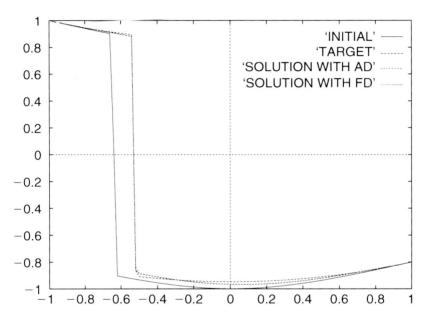

FIG. 5.4. FD vs. AD: solution of the Burgers equation with control in the right-hand side (discretization with 100 intervals). Target, initial and final solutions.

We would like to recover a desired state by solving the following control problem for f:

$$\frac{\partial u}{\partial t} + \frac{\partial}{\partial x}\frac{u^2}{2} = f(x)u, \quad u(-1) = 1, u(1) = -0.8, \tag{5.27}$$

where $f(x) \in \mathcal{O}_{ad}$ is piecewise linear on the mesh, thereby defined by its 100 values of the mesh points.

The cost function is given by:

$$J_\alpha(x) = \frac{1}{2}\int_{-1}^{1}(u(x) - u_d(x))^\alpha dx, \tag{5.28}$$

where $\alpha = 2$ and u_d is the solution obtained solving (5.23) shown in Fig. 5.4. $f(x)$ has been perturbed initially around the target $f_{\text{tar}} = 0.3x$.

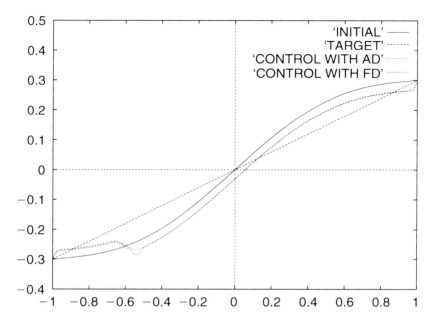

FIG. 5.5. FD vs. AD: control distribution: target, initial and final control state.

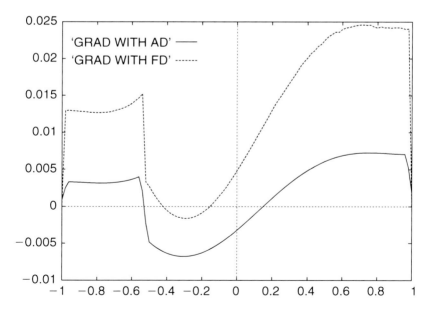

FIG. 5.6. FD vs. AD: the gradients produced by FD and AD approaches.

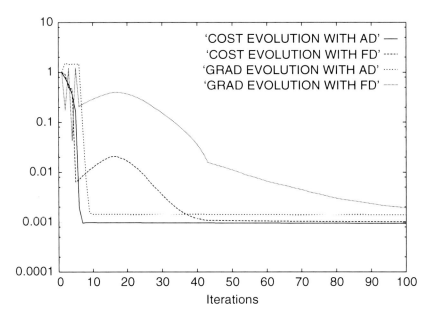

FIG. 5.7. FD vs. AD: the convergence histories for the cost and gradient. When using automatic differentiation, the convergence is uniform. The final states are similar however.

Obviously, it is possible to explicitly calculate f_{tar} knowing $u_d(t,x)$. Indeed, from the state equation in regions where the solution has enough regularity one can write:

$$u_d(t,x) f_{\text{tar}}(x) = \partial_t(u_d) + u_d \partial_x(u_d). \tag{5.29}$$

The control is not a function of time, hence $\partial_x(u_d(t,x)) = 0.3x$. This example points to a fundamental difficulty of inverse design methods as the existence of a solution is not guaranteed in general.

We want to use a gradient method to minimize (5.28) with $\alpha = 2$. The descent parameter is set to 0.001. To get the Jacobian of J, we use AD in reverse mode and finite differences (see Fig. 5.6). We can see that AD improves the convergence (see Fig. 5.7) which means that the gradient is more accurate; however, in both cases, the minimization algorithm fails in finding the global minimum.

The parameterization linked to the discretization is similar to what we will call CAD-free and low-dimensional parameterization is comparable to a CAD-based parameterization of a shape. We come back to the loss of regularity building minimizing sequences using gradients of functionals.

Let us use this example to introduce multicriteria minimization. To see what goes wrong with the convergence of the gradient method we would like to have a geometric view of the functional. Let us consider a low-dimensional search space $\mathcal{O}_{ad}$ to be sampled. For instance, consider $\mathcal{O}_{ad} = \mathcal{P}_1$ a two-dimensional search

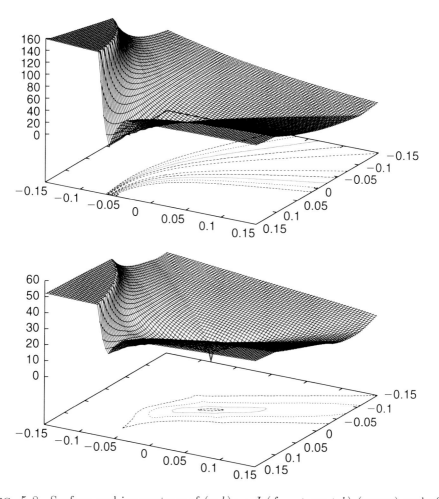

FIG. 5.8. Surface and iso-contour of $(a,b) \mapsto J_2(f_{\text{tar}} + ax + b)$ (upper) and of $(a,b) \mapsto J_{0.3}(f_{\text{tar}} + ax + b)$ (lower).

space of affine polynomials over $[-1, 1]$. With this search space the problem is admissible as $f_{\text{tar}} \in \mathcal{O}_{ad}$.

Figure 5.8 shows:

$$\begin{cases} [-0.15, 0.15] \times [-0.15, 0.15] \to \mathbf{R} \\ (a,b) \mapsto J_2(f_{\text{tar}} + ax + b). \end{cases} \quad (5.30)$$

We can see that the iso-contours are rather parallel around $a = b = 0$ and J_2 is flat in some direction. Let us now consider α as an additional control parameter. One would like to define the best α convexifying the functional. This can be done by dichotomies, for instance, on α. We find that $\alpha = 0.3$ is a suitable choice (see Fig. 5.8).

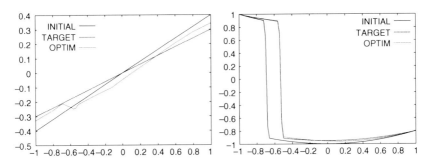

FIG. 5.9. Left: $f_{\text{ini}} = 0.4\, x$, $f_{\text{tar}} = 0.3\, x$ and f_{opt}. Right: u_{ini}, u_{des} and u_{opt} for $\mathcal{O}_{ad} = \mathbf{R}^{100}$. The initial guess is uniform discretization of $0.4x$. One looses regularity.

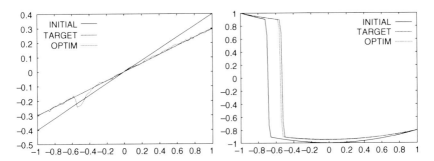

FIG. 5.10. Left: initial $f_{\text{ini}} = 0.4\, x$, target $f_{\text{tar}} = 0.3\, x$ and f_{opt} for $J_{0.3}$. Right: u_{ini}, u_{des} and u_{opt}.

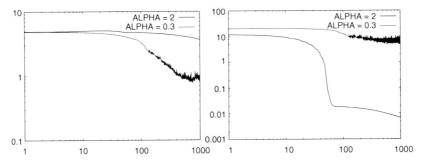

FIG. 5.11. Left: convergence history for $||f - f_{\text{tar}}||$. Right: histories of J_2 and $J_{0.3}$.

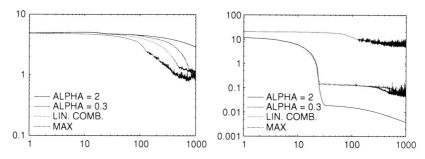

FIG. 5.12. Left: histories of $\|f - f_{\text{tar}}\|$. Right: histories of J_2, $J_{0.3}$, $(0.01\,J_{0.3} + 0.99\,J_2)$ and $\max(J_2, 0.01\,J_{0.3})$.

Now, if one considers again $\mathcal{O}_{ad} = \mathbf{R}^{100}$ with $\alpha = 0.3$ we notice that the convergence is satisfactory in control but less in state than with $\alpha = 2$. It is clear that $J_{\alpha=2}$ and $J_{\alpha=0.3}$ have different behavior and need to be minimized together. This is typical in multicriteria situations. This can be done using a min-max formulation:

$$\min(\max(J_{\alpha=2}, J_{\alpha=0.3})) \tag{5.31}$$

or through a combination of $J_{\alpha=2}$ and $J_{\alpha=0.3}$:

$$\min\left(\frac{(1-\epsilon)}{J^0_{\alpha=2}} J_{\alpha=2} + \frac{\epsilon}{J^0_{\alpha=0.3}} J_{\alpha=0.3}\right), \tag{5.32}$$

with $0 < \epsilon < 1$, $J^0_{\alpha=2}$ and $J^0_{\alpha=0.3}$ being respective initial values of the functionals. One notices that both approaches improve the convergence (see Fig. 5.12).

5.5 Sensitivity in the presence of shocks

As pointed out in the pioneering papers by Majda [32] and Godlewski et al [22] there are serious difficulties in analysis with the calculus of variations when the solution of the partial differential equation has a discontinuity. Optimization of an airplane with respect to its sonic boom, as analyzed below, is precisely a problem in that class, so these difficulties must be investigated. The easiest for us here is to take a simple example, like the Burgers equation, to expose the problem and the results known so far.

Suppose we seek the minimum with respect to a parameter a (scalar for clarity) of $j(u,a)$ with u a solution of

$$\partial_t u(x,t) + \partial_x\left(\frac{u^2}{2}\right)(x,t) = 0, \quad u(x,0) = u^0(x,a), \quad \forall (x,t) \in R \times (0,T). \tag{5.33}$$

Consider initial data u^0, with a discontinuity at $x = 0$ satisfying the entropy condition $u^-(0) > u^+(0)$; then $u(x,t)$ has a discontinuity at $x = s(t)$ which depends on a of course, and propagates at a velocity given by the Rankine-Hugoniot condition: $\dot s = \bar u := (u^+ + u^-)/2$ where $u^\pm$ denote its values before and after the shock.

Let H denotes the Heaviside function and δ its derivative, the Dirac function; let $s' = \partial s/\partial a$ and $[u] = u^+ - u^-$ the jump of u across the shock. We have

$$u(x,t) = u^-(x,t) + (u^+(x,t) - u^-(x,t))H(x - s(t)) \quad \Rightarrow$$
$$u' = u^{-'} - s'(t)[u]\delta(x - s(t)), \tag{5.34}$$

where $u^{-'}$ is the pointwise derivative of u^- with respect to a. One would like to write that (5.33) implies

$$\partial_t u'(x,t) + \partial_x(uu')(x,t) = 0, \qquad u'(x,0) = u^{0'}(x,a). \tag{5.35}$$

Unfortunately uu' in (5.35) has no meaning at $s(t)$ because it involves the product of a Dirac function with a discontinuous function! The classical solution to this riddle is to say that (5.34) is valid at all points except at $(t, s(t))$, and that the Rankine-Hugoniot condition, differentiated, gives

$$\dot{s}'(t) = \bar{u}'(s(t),t) + s'(t)\partial_x \bar{u}(t, s(t)). \tag{5.36}$$

Here we see that the derivative of the Burgers equation has two unknowns: u' and s'. The entropy condition insures uniqueness of u' given by (5.35) and a jump condition across the shock is not necessary because the characteristics left and right of the shock point left and right also. Once u' is computed, s' is given by the linear ODE (5.36).

The question then is to embed these results into a variational framework so as to compute the derivative of j as usual by using weak forms of the PDEs and adjoint states. It turns out [4] that (5.35) is true even at the shock, but in the sense of distribution theory and with the convention that whenever uu' occurs it means $\bar{u}u'$ at the shock. Furthermore (5.35) in the sense of distribution contains a jump condition which, of course, is (5.36). This apparently technical result has a useful corollary: integrations by parts are valid, so the real meaning of (5.35) is

$$\int_R (u'(T)w(T) - u^{0'}w(0)) - \int_{R\times(0,T)} u'(\partial_t w + u\partial_x w) = 0 \quad \forall w. \tag{5.37}$$

and it has a unique (distribution) solution. Of course, the derivative of $j = \int_{R\times(0,T)} J(x,t,u,a)$ with respect to a is $j' = \int_{R\times(0,T)} (J'_a + J'_u u')$.

When a is multidimensional, to transform $\int_{R\times(0,T)} J'_u u'$, one may introduce an adjoint state v solution of

$$\partial_t v + u\partial_x v = J'_u(x,t), \qquad v(x,T) = 0, \tag{5.38}$$

and write (see (5.37))

$$\int_{R\times(0,T)} J'_u u' = \int_{R\times(0,T)} (\partial_t v + \bar{u}\partial_x v)u' = -\int_R u^{0'} v(0). \tag{5.39}$$

Notice that the adjoint state v has no shock because its time boundary condition is continuous and the characteristics integrated backward never cross the shock.

This fact was observed by Giles [20, 21] in the more general context of the Euler equations for a perfect gas.

This analysis shows that a blind calculus of variation is valid, so long as the criterion j does not involve the position of the shock or its amplitude explicitly and that there is no need to include in a calculus of sensitivity the variations of the shock position explicitly.

5.6 A shock problem solved by AD

We consider the problem of computing the sensitivity of the shock position in a transonic nozzle. Consider the Euler equations:

$$\partial_t W + \nabla \cdot F(W) = 0, \quad W = \begin{pmatrix} \rho \\ \rho \vec{u} \\ \rho E \end{pmatrix}, \quad (5.40)$$

With an entropy condition, initial conditions, and boundary conditions such as $W \cdot n|_{in} = g(\alpha)$.

Consider now the problem of computing the derivative V of W with respect to a parameter α appearing in the boundary conditions [21]. As calculus of variation applies we would expect V to be the solution of the linearized Euler equations

$$\partial_t V + \nabla \cdot F'(W)V = 0, \quad V = \begin{pmatrix} \rho' \\ (\rho \vec{u})' \\ (\rho E)' \end{pmatrix}, \quad (5.41)$$

with boundary condition $V \cdot n|_{in} = g'(\alpha)$. However, if W has a shock, we know that the shock position depends upon α. But how can $W + V\delta\alpha$ displace the shock?

Above, we saw the analysis for the Burgers equation $(f(u) = u^2/2)$.

$$\partial_t u + \partial_x f(u) = 0, \quad (5.42)$$

and that the linearized Burgers equation in weak form

$$\langle \partial_t w, v \rangle + \langle \partial_x w, f'(u)v \rangle = 0 \quad \forall w \in C^1(R \times R^+), \quad (5.43)$$

makes sense provided that $f'(u)v$ which is the product of a discontinuous function by a distribution means $\frac{1}{2}(f'(u)^+ + f'(u)^-)v$ which, as we have seen, contains the Dirac mass $[f'(u)]\delta_\Sigma$ at the shock position Σ.

Figure 5.13 displays two computations for two different values of the inlet Mach number for a nonviscous flow. Automatic mesh adaptation by metric control has been used for accurate results.

The first case was then redone with AD. Figure 5.6 displays the derivative of the Mach number and of the density on the symmetry line (bottom line here) of the nozzle around the shock location.

Both display a Dirac mass at the shock position.

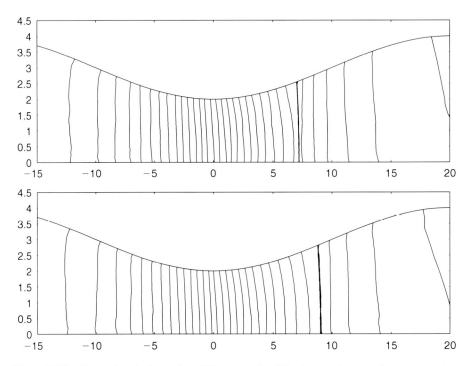

FIG. 5.13. Two simulations for different inlet Mach numbers; adaptive meshes were used so as to resolve the shocks properly.

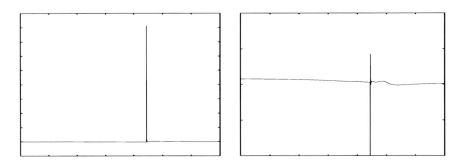

FIG. 5.14. Derivatives of the local Mach number (left) and local density (right) on the symmetry line of the nozzle around the shock location.

5.7 Adjoint variable and mesh adaptation

One might be interested in optimizing or adapting a mesh not for the whole state calculation but for the evaluation of a particular functional. It is likely that the overall mesh size will be less if less details or information are required for the evaluation of this functional. For instance, one might be interested in computing the drag or lift coefficient with some given precision without being interested in all the details of the flow away from the aircraft. One would like to use the concepts introduced in Chapter 4 for mesh adaptation by metric control together with the adjoint variable calculation [5, 8, 19, 40, 50].

Let us consider the problem of computing a functional $J_h(x, u_h)$ with u_h a discrete state solution of a discrete state equation $F_h(x, u_h(x)) = 0$, on a mesh of size h where x are the optimization parameters. For the sake of simplicity, without loss of generality, let us assume that J is computable exactly, i.e. $J = J_h$.

One has in addition the solution of a discrete adjoint problem:

$$v_h^T \left(\frac{\partial F_h}{\partial u_h} \right) = \frac{\partial J}{\partial u_h}, \qquad (5.44)$$

which gives the gradient of the functional as:

$$\frac{dJ}{dx} = \frac{\partial J}{\partial x} + v_h^T \frac{\partial F_h}{\partial x}. \qquad (5.45)$$

Our aim is to take advantage of the knowledge of v_h to find a mesh optimal for the calculation of J minimizing the error on J: $\varepsilon_J = |J(u) - J(u_h)|$ when u and u_h satisfy $F(u) = 0$ and $F_h(u_h)$.

The error on J can be linked to the error on the state δu_h (i.e. $u = u_h + \delta u_h$) by

$$J(u) - J(u_h) \approx \frac{\partial J}{\partial u_h} \delta u_h \approx \frac{\partial J}{\partial u_h} \left(\frac{\partial F_h}{\partial u_h} \right)^{-1} F_h(u) = v_h^T F_h(u), \qquad (5.46)$$

because

$$F_h(u) \approx F_h(u_h) + \frac{\partial F_h}{\partial u_h} \delta u_h = \frac{\partial F_h}{\partial u_h} \delta u_h, \qquad (5.47)$$

and $F_h(u_h) = 0$.

Hence, if one has an estimation of the error $F_h(u)$ one can deduce an estimation of the error on the functional. In case such estimation is not available, it is possible to build one, from an estimate like

$$\|F_h(u)\| \leq Ch^p \|\nabla^q u\|. \qquad (5.48)$$

The constant C and powers p and q shall be estimated using convergence diagrams from calculations on successive (say three) refined meshes for a case where the analytical solution u is known. This final limitation can be removed considering the same iterative adaptation spirit replacing u by u_h. One can now introduce ε_J in the expression seen in (4.60) used to equidistribute the interpolation error by defining a uniform mesh in a metric M (a_i being a segment in an unstructured mesh):

$$\frac{1}{c \mathcal{E} \varepsilon_J} a_i^T M a_i = 1. \qquad (5.49)$$

Hence, one can afford longer edges and therefore coarser meshes if the error in the functional is small. This approach is also useful to provide information on a fine level mesh of size h from calculations of the state and adjoint on a coarse mesh of size H by looking for errors such as $\varepsilon_J = |J(u_h) - J(u_H)|$ where u_h is out of reach while u_H and v_H can be obtained at reasonable cost.

5.8 Tapenade

The program used here for computing the gradient of the cost function has been obtained using the automatic differentiator `Tapenade` developed at INRIA-Sophia Antipolis by L. Hascoet and his team Tropics following the initial development of `Odyssée` [18, 43, 17, 23]. This tool reads standard `Fortran` 7777 or 95 or C programs. One important feature is that non-differentiable predefined functions like *min, max, sign, abs,* etc. can be treated. Inter-procedural derivation is also possible.

Two differentiation procedures are available, called the forward and the reverse modes.

The forward or direct mode consists of computing the function and its derivative at the same time. When using the direct mode, the user has to choose between three algorithms depending on the nature of the performance required. One can compute the function and all the partial derivatives at the same time. This is the most memory consuming choice. Alternatively, one can compute the function and one partial derivative each time. This is quite like using finite differences. This choice is the most time consuming because of the redundant evaluations of the function. Finally, one can compute the function and save the computational graph of the function, then compute the partial derivatives using the dependency information from the graph. This choice might be quite memory consuming if the graph is complex.

An estimation of the time needed by the second algorithm for programs where only arithmetic operations are involved is given by:

$$T(f, f') \leq 4nT(f), \tag{5.50}$$

where $T(f)$ is the time for an evaluation of the function and n is the number of control points. We can see that this is more than for finite differences.

The backward mode can be seen as the adjoint method for computing gradients of optimization problems. The most important advantage of this approach compared to the previous one is that $T(f, f')/T(f)$ is bounded by a constant independent of the number of control points.

5.9 Direct and reverse modes of AD

We give an alternative presentation of these modes already introduced at the beginning of the chapter. Let f be a composed function given by:

$$x \in R^p \to y = h(x) \in R^n \to z = g(y) \in R^n \to u = f(z) \in R^q. \tag{5.51}$$

Following the composed function differentiation rule we have a product of matrices.

$$u' = f'(z)g'(y)h'(x), \tag{5.52}$$

where $f' \in R^{q \times n}$, $g' \in R^{n \times n}$, $h' \in R^{n \times p}$. We observe that from a practical point of view, in (5.52) we need to introduce an intermediate matrix $M = g'(y)h'(x) \in R^{p \times n}$ to store the intermediate result before making $u' = f'(z)M$.

Now, after transposition of (5.52) we have:

$$u'^T = h'^T(x)g'^T(y)f'^T(z). \tag{5.53}$$

We can see that the storage is now $M = g'^T(y)f'^T(z) \in R^{n \times q}$. It is easy to understand that according to the dimensions of the different spaces (i.e. p and q), we should use formula (5.52) or (5.53) to optimize the required memory. For instance, for optimization applications where p is the number of control variables and $q = 1$ with f being a cost function, differentiation after transposition is more suitable.

We call the choice (5.52) the direct and (5.53) the reverse mode of differentiation.

In a computer program, the situation is the same. In the direct mode, the Jacobian is produced by differentiating the program (considered as a composed function) line by line. The reverse mode is less intuitive, it corresponds to writing the adjoint code (instructions of the direct code in the reverse order). We will describe these modes through examples.

Consider the following function $f = x^2 + 3x$ ($f' = 2x + 3$), written as a composed function:

```
y_1=x
y_2=x**2+2*y_1
f  =y_1+y_2
```

We are looking for the derivative of f with respect of x.

Using the direct mode   A line by line derivation with respect of x will give:

$$\frac{dy_1}{dx} = 1, \quad \frac{dy_2}{dx} = 2x + 2\frac{dy_1}{dx}, \tag{5.54}$$

$$\frac{df}{dx} = \frac{dy_1}{dx} + \frac{dy_2}{dx} = 1 + 2x + 2. \tag{5.55}$$

We see that we have to store all the intermediate computation before making the final addition.

Using the reverse mode   Here, we consider f as a cost function and the lines of the program as constraints. Hence, we can define an augmented Lagrangian for this program associating one Lagrange multiplier to each affectation:

$$L = y_1 + y_2 + p_1(y_1 - x) + p_2(y_2 - x^2 - 2y_1). \tag{5.56}$$

The Jacobian is a saddle-point for this Lagrangian. On the other hand, the derivative of the Lagrangian with respect to intermediate variables is zero and with respect of the independent variable is the Jacobian of f:

$$\frac{df}{dx} = \frac{\partial L}{\partial x} = -p_1 - 2p_2 x,$$
$$\frac{\partial L}{\partial y_1} = 1 + p_1 - 2p_2 = 0,$$
$$\frac{\partial L}{\partial y_2} = 1 + p_2 = 0. \tag{5.57}$$

We notice that to get the Jacobian df/dx, the previous equations have to be solved in reverse order through an upper triangular system. We always have an upper triangular system due to the fact that in an affectation we have only one entity in the left-hand side. This presentation of the reverse mode is quite elegant but not easy to implement. In Tapenade differentiator tool the adjoint code method, presented below, has been used

The adjoint code method The idea is to write the adjoint of the direct code line by line taken in reverse order. The key is that for each affectation like $y = y + f(x)$, the dual expression is $p_x = p_x + f' p_y$ with p_x and p_y the dual variables associated to x and y. The previous example becomes:

$$p_{y_1} = p_{y_2} = p_x = 0, \quad p_f = 1, \tag{5.58}$$
$$p_{y_1} = p_{y_1} + p_f = 1, \quad p_{y_2} = p_{y_2} + p_f = 1, \tag{5.59}$$
$$p_{y_1} = p_{y_1} + 2p_{y_2} = 3, \quad p_x = p_x + 2xp_{y_2} = 2x, \tag{5.60}$$
$$p_x = p_x + p_{y_1} = 2x + 3. \tag{5.61}$$

We can see that no triangular system is solved or stored and that the Lagrangian has not been formed.

do - if

The most frequent instructions in finite element, volume or difference solvers are loops and conditional operations (often hidden through abs, min, max, etc.). We present an example to explain how these are treated with AD.

Consider the evolution of $|u(t)|$ (nondifferentiable) with respect of the initial condition u_0, u being the solution of:

$$\frac{du}{dt} = -au, \quad u(0) = u_0. \tag{5.62}$$

We use an explicit discretization:

$$\frac{u^{i+1} - u^i}{\delta t} = -au^i, \tag{5.63}$$

which can be programmed as:

$$u = u_0$$

$$\begin{aligned}&\text{do } i = 1, ..., N\\&v = -au\\&u = u + \delta t\, v\\&\text{enddo}\\&f = |u|.\end{aligned} \qquad (5.64)$$

After expansion:

$$\begin{aligned}&u_1 = u_0, v_1 = -au_1, u_2 = u_1 + \delta t v_1,\\&v_2 = -au_2, ..., v_N = -au_{N-1}, u_{N+1} = u_N + \delta t v_N,\end{aligned} \qquad (5.65)$$

we introduce the Lagrangian of the program as before:

$$L = |u_{N+1}| + p_0(u_1 - u_0) + \sum_{i=1}^{N}(p_i(v_i + au_i) + p'_i(u_{i+1} - u_i + \delta t v_i)). \qquad (5.66)$$

Optimality conditions give:

$$\begin{aligned}\frac{\partial L}{\partial u_0} &= \frac{\partial f}{\partial u_0} = -p_0,\\\frac{\partial L}{\partial u_1} &= p_0 + p_1 a - p'_1,\\\frac{\partial L}{\partial v_i} &= p_i + p'_i \delta t, \quad i = 1, ..., N\\\frac{\partial L}{\partial u_i} &= p_i a - p'_i, \quad i = 1, ..., N\\if(u < 0) \quad \frac{\partial L}{\partial u_{N+1}} &= -1 + p'_N,\\if(u \geq 0) \quad \frac{\partial L}{\partial u_{N+1}} &= 1 + p'_N.\end{aligned} \qquad (5.67)$$

The limit of the method is the memory required to store p_i and p'_i, especially if internal loops are present. We can see that the branches of conditional statements are treated separately and that the results are assembled after derivation.

5.10 More on FAD classes

Derivatives of functions can be computed exactly not only by hand but also by computers. Commercial software such as Maple [52] or Mathematica [53] have derivation operators implemented by formal calculus techniques. In [24] Griewank presented a C++ implementation using operator overloading, called Adol-C. Speelpenning's tool Jake-f also allowed insertion of subroutines in order to perform formal calculus on each instruction of a computer program. Thus a function described by its computer implementation can be differentiated *exactly* and *automatically*. This is the reason why it is now called automatic differentiation of computational approximations to functions [16].

The idea of using operator overloading for AD can be traced to [6], [24], and [9]. It has been used extensively in [25] for the computation of Taylor series of computer functions automatically and we wish to acknowledge the fact that it is this later work which has instigated this study.

Expression templates were introduced by T. Veldhuizen [49] in 1995 for vectors and array computations. Using these techniques, he provided an *active library* called `Blitz++` (http://monet.uwaterloo.ca/blitz) that enables `Fortran 77`performance for C++ programs on several Unix platforms (IBM, HP, SGI, LINUX workstations and Cray T3E). Expression templates avoid the creation of temporary arrays

for y = A*x,   C++ compilers do   c←A*x, c1←c, y←c1

Traits were introduced by N. Myers [38] for type promotion for templates classes. We consider the addition of two `Fad` of different types :

Fad<TYPE> = Fad<double> + Fad<std::complex<float> >

The problem is to know automatically the return type TYPE. For example C's promotion rules and mathematical promotion rules can be used in simple cases:

C rules : float + double →double,   int + float → float, ...
Mathematical rules : double + complex →complex, ...

and we apply the rules to `Fad<>` calculation. But to avoid the creation of a temporary location to store the matrix-vector product Ax an automatic generator of the type `MatrixVectorProductResult` must be generated and that is done with traits.

P. Aubert and N. Dicesare wrote `Fad.h` which implements and tests these ideas. `Adol-C`(Griewank), `Fadbad` 2.0 (Bendtsen and Stauming) and `Fad` (Aubert and Dicesare), with and without expression templates, and analysis shows that it is equal to analytic methods that have been compared (see Fig. 5.15).

In order to provide a more understandable test, computations have been done with the forward mode of `Adol-C1.8` [24] and `Fadbad` 2.0 because `Adol-C`required several changes to be compilable with the KCC (see Fig. 5.16). `Adol-C`also tries to minimize the number of temporaries and loops introduced by the classical overloading technique. But it is managed using pointers and not auxiliary template classes. `Fadbad` uses the classical overloading approach.

```
// Adol-C trace_on(1);
y = 0.;
for(i=0; i<n; i++) {
   tmp <<= xp[i];
   y = ((y*tmp) + (tmp/tmp)) -
   ((tmp*tmp) + (tmp - tmp));
```

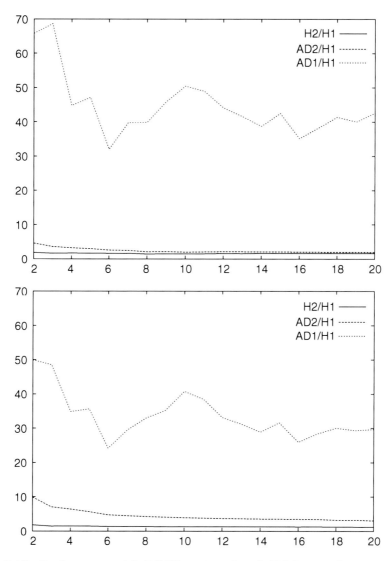

FIG. 5.15. Performance with EGCS 1.1.2 and with KCC 3.3g: Comparison between by hand (H), by overloading without expression templates and with expression templates. The number variables goes from 1 to 20. We compute nloop times 2.f*(x1*x2)-x3+x3/x5+4.f.

```
} y >>= yp; trace_off();
forward(1,1,n,0,xp,g);// gradient
evaluation
```

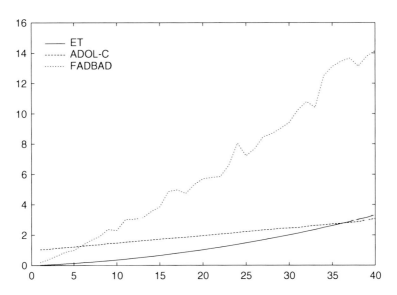

FIG. 5.16. Adol-C1.8, Fadbad 2.0 and the Fad<> class comparison.

```
// FadBad Fdouble gradients and values
// computed at the same time
y = 0.; for(i=0; i<n; i++) {
  Fdouble tmp(xp[i]);
  tmp.diff(i,n);
  y = ((y*tmp) + (tmp/tmp)) -
  ((tmp*tmp) + (tmp - tmp));
}

// Fad<> gradients and values computed
// at the same time
y = 0.;
for(i=0; i<n; i++) {
  Fad<double> tmp(xp[i]);
  tmp.diff(i,n);
  y = ((y*tmp) + (tmp/tmp)) -
  ((tmp*tmp) + (tmp - tmp));  }
```

The test defines a vector of *independent* variables and performs several arithmetic operations on this vector that are accumulated in a single variable y. The test computes the derivatives of y with respect to the *independent* variables. Figure 5.16 is a plot of the computation times with respect to the number of *independent* variables.

In Fig. 5.16, the method using expression templates (ET) is clearly the fastest until the number of *independent* variables is greater than 50. But it is beyond the scope of the forward mode. In comparison to `Adol-C`, ET has a quadratic growth. This could be explained by the poor inlining capabilities of EGCS. A comparison with KCC would be instructive but `Adol-C`does not compile with KCC at the moment.

References

[1] Anderson, K. Newman, J. Whitfield, D. and Nielsen, E. (1999). Sensitivity analysis for the Navier-Stokes equations on unstructured grids using complex variables, *American Institute of Aeronautics and Astronautics*, **99-329**.

[2] Aubert, P. (2000). Cross-section optimisation for curved and twisted 3D beams via a mixed forward and backward optimisation algorithm, *Proc. AD 2000 SIAM conference*, Nice.

[3] Aubert, P. Dicesare, N. and Pironneau, O. (1999). *Automatic Differentiation in C++ using Expression Templates and Application to a Flow Control Problem,* Computer Visualization and Software, Springer, Berlin.

[4] Bardos, C. and Pironneau, O. (2002). A formalism for the differentiation of conservation laws. *C. R. Acad. Sci., Paris, Serie I*. Numerical Analysis, **453**, 839-845.

[5] Baker, T.J. (1997). Mesh adaptation strategies for problems in fluid dynamics, *Finite Element Anal. Design*, **25**, 243273.

[6] Barton, J. and Nackman, L. (1994). *Scientific and Engineering C++,* Addison-Wesley.

[7] Baysal, O. and Eleshaky, M.E. (1992). Aerodynamic design optimization using sensitivity analysis and CFD, *Amer. Inst. Aeronaut. Astronaut. J.* **30(3)**, 718-725.

[8] Becker, R. and Rannacher, R. (2001). An optimal control approach to a posteriori error estimation in finite element methods, *Acta Numerica*, **10**, 50-102.

[9] Bendtsen, C. and Stauning, O. (1996). `Fadbad`, a fexible C++ package for automatic differentiation using the forward and backward methods, *Technical Report* IMM-REP-1996-17.

[10] Cabot, A. and Mohammadi, B. (2002). Incomplete sensitivities and cost function reformulation leading to multicriteria investigation of inverse problems *Opt. Contr. and Meth.* **24(2)**, 73-84.

[11] Carpentieri, G. Van Tooren, M. and Koren, B. (2006). Improving the efficiency of aerodynamic shape optimization on unstructured meshes. *AIAA*, **2006-298**.

[12] Carpentieri, G. Van Tooren, M. and Koren, B.(2007). Development of the discrete adjoint for a 3D unstructured Euler solver. *AIAA*, **2007-3954.**

[13] Brenier, Y. Osher, S. (1986). Approximate Riemann solvers and numerical flux functions. *SINUM,* **23**, 259-273.

[14] Brown, P. and Saad, Y. (1990). Hybrid Krylov methods for nonlinear systems of equations, *SIAM J. Sci. Stat. Comp.*, **11(3)**, 450-481.

[15] Courty, F. Dervieux, A. Koobus, B. and L. Hascoët. (2001). Reverse automatic differentiation for optimum design: from adjoint state assembly to gradient computation, *Optim. Meth. Software,* **18(5)**, 615-627.

[16] Eberhard, P. (1999). Argonne theory institute: differentiation of computational approximations to functions, *SIAM News*, **32(1)**, 148-149.

[17] Faure, C. (1996). Splitting of algebraic expressions for automatic differentiation, *Proc. of the Second SIAM Inter. Workshop on Computational Differentiation,* Santa Fe.

[18] Gilbert, J.C. Le Vey, G. and Masse, J. (1991). La différentiation automatique de fonctions représentées par des programmes, *INRIA* **RR-1557**.

[19] Giles, M. (1998). *On Adjoint Equations for Error Analysis and Optimal Grid Adaptation in CFD,* in Computing the Future II: Advances and Prospects in Computational Aerodynamics, edited by M. Hafez and D. A. Caughey, John Wiley, New York.

[20] Giles, M. and Pierce, N. (1997). Adjoint equations in CFD: duality, boundary conditions and solution behaviour, *American Institute of Aeronautics and Astronautics* **97-1850**.

[21] Giles, M. and Pierce, N. (2001). Analytic adjoint solutions for the quasi-one-dimensional euler equations. *J. Fluid Mech.,* **426**, 327-345.

[22] Godlewski, E. Olazabal, M. and Raviart, P.A. (1998). *On the Linearization of Hyperbolic Systems of Conservation Laws,* Eq. aux dérivées partielles et applications, Gauthier-Villars, Paris.

[23] Griewank, A. (2001). *Computational derivatives,* Springer, New York.

[24] Griewank, A. Juedes, D. and Utke, J. (1996). Algorithm 755: ADOL-C: a package for the automatic differentiation of algorithms written in C/ C++, *J-TOMS,* **22(2)**, 131-167.

[25] Guillaume, Ph. and Masmoudi, M. (1997). Solution to the time-harmonic Maxwell's equations in a waveguide: use of higher-order derivatives for solving the discrete problem, *SIAM J. Numer. Anal.,* **34(4)**, 1306-1330.

[26] Hafez, M. Mohammadi, B. and Pironneau, O. (1996). Optimum shape design using automatic differentiation in reverse mode, *Int. Conf. Num. Meth. Fluid Dyn.,* Monterey.

[27] Hansen, L.P. (1982). Large sample properties of generalized method of moments estimators, *Econometrica*, **50**, 1029-1054.

[28] Jameson, A. (1994). Optimum aerodynamic design via boundary control, *AGARD Report* **803**, Von Karman Institute Courses.

[29] Launder, B.E. and Spalding, D.B. (1972).*Mathematical Models of Turbulence,* Academic Press, London.

[30] Madsen, H. O., Krenk, S., and Lind, N.C.(1986). *Methods of Structural Safety*, Prentice-Hall, Englewood Cliffs, NJ.

[31] Melchers, R.E. (1999). *Structural Reliability Analysis and Prediction*, 2nd edition, John Wiley, London.

[32] Majda, A. (1983). The stability of multi-dimensional shock fronts. *Memoire of the A.M.S.* 281, American Math. Soc. Providence.

[33] Makinen, R. (1999). Combining automatic derivatives and hand-coded derivatives in sensitivity analysis for shape optimization problems. *Proc. WCSMO 3*, Buffalo, NY.

[34] Mavriplis, D. (2006). A discrete adjoint-based approach for optimization problems on three-dimensional unstructured meshes. *AIAA*, **2006-50**.

[35] Mohammadi, B. (1996). Optimal shape design, reverse mode of automatic differentiation and turbulence, *AIAA*, **97-0099**.

[36] Mohammadi, B. (1996). Différentiation automatique par programme et optimisation de formes aérodynamiques, *Matapli* **07-96**, 30-35.

[37] Mohammadi, B. Malé, J.M. and Rostaing-Schmidt, N. (1995). Automatic differentiation in direct and reverse modes: application to optimum shapes design in fluid mechanics, *Proc. SIAM workshop on AD*, Santa Fe.

[38] Myers, N. (1995). A new and useful template technique: traits, *C++ Report*, **7(5)**, 32-35.

[39] Papadimitriou, D. and Giannakoglou, K. (2006). Continuous adjoint method for the minimization of losses in cascade viscous flows. *AIAA*, **2006-49**.

[40] Pierce, N. and Giles, M. (2000). Adjoint recovery of superconvergent functionals from PDE approximations, *SIAM Rev.* **42(247)**.

[41] Pironneau, O. (1984).*Optimal Shape Design for Elliptic Systems,* Springer-Verlag, New York.

[42] Roe, PL. (1981). Approximate Riemann solvers, parameters vectors and difference schemes, *J. Comp. Phys.* **43**, 357372.

[43] Rostaing, N. (1993). *Différentiation automatique: application à un problème d'optimisation en météorologie*, Thesis, University of Nice.

[44] Rostaing, N. Dalmas, S. and Galligo. A. (1993). Automatic differentiation in Odyssée, *Tellus*, **45a(5)**, 558-568.

[45] Squire, W. and Trapp, G. (1998). Using complex variables to estimate derivatives of real functions , *SIAM review*, **10(1)**, 110-112.

[46] Stroustrup, B. (1997). *The C++ Programming Language,* Addison-Wesley, New York.

[47] Su, J. and Renaud, J.E. (2001). Automatic Differentiation in robust optimization, *AIAA J.*, **35:6**.

[48] Van Albada G.D. and Van Leer, B. (1984), Flux vector splitting and Runge Kutta methods for the Euler equations, *ICASE Report* **84-27**.

[49] Veldhuizen, T. L. (1995) Expression templates, *C++ Report,* **7(5)** 26:31, Reprinted in C++ Gems, ed. Stanley Lippman.

[50] Venditti, D. A. and Darmofal, D. L. (2000), Adjoint error estimation and grid adaptation for functional outputs, *JCP*, **164**, 204227.

[51] Viollet, P. L. (1981). On the modelling of turbulent heat and mass transfers for computation of buyoancy affected ows. *Proc. Int. Numer. Meth. for Laminar and Turbulent Flows*, Venezia.

[52] Waterloo Maple software. (1995). *Maple Manual.*

[53] Wolfram Research. (1995). *Mathematica Manual.*

6
PARAMETERIZATION AND IMPLEMENTATION ISSUES

6.1 Introduction

In this chapter we briefly describe other ingredients needed to build a design and control platform. These are:

- The definition of a control space (shape parameterization).
- Operators acting on elements of the control space (shape deformation tool).
- Operators to link the control space and the geometrical entities (mesh deformation tool).
- Geometric and state constraint definition.
- How to link shape optimization and mesh adaptation.

For each item, we will only describe the approach we use currently ourselves. Therefore the descriptions are not exhaustive. The idea is to show the kind of results that can be obtained with a platform built with these tools. We are interested in particular with gradient-based minimization algorithms which have a pseudo-unsteady interpretation.

6.2 Shape parameterization and deformation

We describe the set of tools needed to define the shape deformation (first arrow below) from a variation of control parameters $x \in \mathbf{R}^n$:

$$\delta x \to \delta x_w \to \delta x_m,$$

where $x_w \in \mathbf{R}^{n_w}$ denotes the set of discretization points on the geometry and $x_m \in \mathbf{R}^N$ the internal mesh nodes.

In the context of shape optimization, control parameters can be for instance a relaxed characteristic function (level set and immersed boundary approaches are in this class), a body fitted parameterization using CAD or a CAD-free parameterization. One can use parametric curves or look for best shape in the convex hull of a set of shapes. Some of these parameterizations and also topological optimization have been mentioned in Chapter 2.

To give an idea of the complexity in terms of space dimension for each of these parameterizations, in a regular 3D mesh characterized by a one-dimensional parameter n, the size of a CAD-free parameter space is $O(n^2)$, between low-dimension CAD type parameter spaces with size $O(n)$ and parameter spaces related to the computational mesh with size in $O(n^3)$. The different approaches should be seen therefore as complementary for the primary and final stages of optimization. A minimization algorithm can be seen as a governing equation for the motion in parameter space and for discussing global optimization issues in Chapter 7, we will see that the same equation can be used for the motion of the level set function, for instance, and for minimization. We also need to monitor the regularity in parameter space during optimization for CAD-free, level set or immersed boundary methods.

Also the choice of parameter shape is made following the variety of the shapes one would like to reach. For instance, if the topology of the target shape is already known and if the available CAD parameter space is thought to be rich enough, it should be considered as a control parameter space during optimization. On the other hand, one might use a different parameter space, larger or smaller, during optimization, having in mind that the final shape should be expressed in a CAD format to be industrially usable.

Below we briefly describe some shape parameterizations.

6.2.1 *Deformation parameterization*

For a given parameterized curve $\gamma(t)$ in $\mathbf{R}^2$ of parameter t, a natural way to specify deformation is to relate these deformations to the local normal $n(t)$ of $\gamma(t)$. Hence, the deformed curve $\tilde{\gamma}$ is

$$\tilde{\gamma}(\alpha, t) = \gamma(t) + f(\alpha, t).n(\gamma(t)).$$

where $x = \alpha \in P \subset \mathbf{R}^n$ belongs to the control space P which is usually aimed at being small.

Another popular approach used for parameterization of deformations in the design for aeronautical problems is based on multi-scale global trigonometric functions defined over the shape [10]. In all these approaches, not the shape but only the deformation is parameterized.

6.2.2 *CAD-based*

In this parameterization the shape and therefore its deformations are defined through a few control points compared to the number of nodes needed for the discretization of the shape for a simulation, with finite elements, for instance:

$$n \ll n_w.$$

Of course, shape deformations should be admissible for the parameterization. For instance, a cubic spline will not allow a singularity in the shape. Another feature of this parameterization is smooth variations of control points when propagating to body discretization points.

6.2.3 *Based on a set of reference shapes*

We can express the deformations as a linear combination of admissible deformations available in a data base. This is probably the most convenient way but it is specific to each application and requires a rich database:

$$S = \Sigma \lambda_i S_i, \quad \Sigma \lambda_i = 1, \quad S \in \{S_i, i = 1, ..., n\}.$$

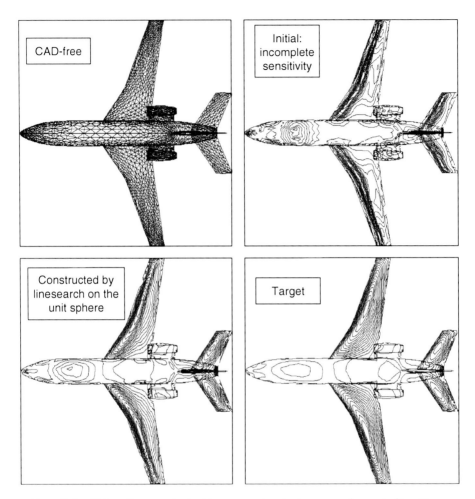

FIG. 6.1. Using the regularization operator to improve descent directions.

6.2.4 *CAD-free*

In this parameterization, we choose to use the same parameterization for the simulation and optimization. This is the simplest explicit shape parameterization: all the nodes of the surface mesh over the shape are control parameters:

$$n = n_w.$$

One feature of this parameterization comes from the fact that, unlike in a CAD-based parameter space, regularity requirements have to be specified and handled by the user. Indeed, if the shape is described using a CAD tool and if we use the same parameterization to specify the deformations, the two entities belong to the same space in terms of regularity.

Hence, the geometrical modeler is removed from the optimization loop. The geometrical modeler is usually quite complicated to differentiate and the source code is not available to the user when a commercial tool is used. Moreover, the CAD parameterization is not necessarily the best for optimization. The idea therefore is to perform the optimization at the h (discretization) level, i.e. on the discrete problem directly. Of course, we have a correspondence between the surface mesh and the CAD parameterization. In this parameterization the only entity known during optimization is the mesh.

We can show on a simple example that the same optimization problem can be easier or harder to solve depending on the parameterization. This is an inverse design problem where one looks at finding a profile x realizing a given state distribution u_d, the solution of a potential flow, minimizing $J(x) = \|u(x) - u_d\|$; u_d is the target field obtained with a cosine shape. One observes that with a CAD type parameterization the convergence is more difficult than with a CAD-free parameterization (see Fig. 6.2). The conditioning of the problem is less favorable with the first parameterization. Therefore, even though the size of the problem is 6 instead of 100 for the CAD-free parameters, the total work appears equivalent if one looks for full convergence. One therefore needs a better performing optimizer to reduce the work with CAD parameters. We again insist on the fact that the final shapes have to be specified in CAD for industrial efficiency even if another parameterization is used during optimization.

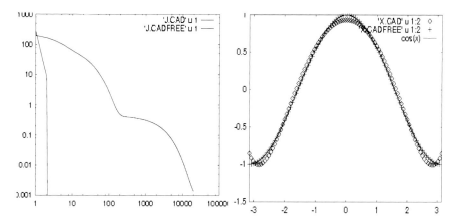

FIG. 6.2. Inverse design with a target shape defined as $\cos(x)$. Left: effect of the parameterization on the convergence of a quasi-Newton method. A CAD parameterization with 6 degrees of freedom versus a CAD-free parameterization based on 100 equidistant nodes $\{x_i, y_i\}$. On the right one sees that in the case of partial convergence (after 100 iterations), small details of the shape are not recovered with the CAD-based solution.

A local second-order smoother This is an attempt to recover the smoothing feature of the CAD-based parameterization when using a CAD-free parameterization. The need for a smoothing operator can find theoretical justification from the consistent approximation theory given in Chapter 6.

The importance of a smoothing step can also be understood by the following argument: If Γ denotes a manifold of dimension $(n-1)$ in a domain $\Omega \in R^n$, we want the variation $\delta x_w \in C^1(\Gamma)$. From Sobolev inclusions, we know that $H^{(2n-1)/2}(\Gamma) \subset C^1(\Gamma)$. It is easy to understand that the gradient method we use does not necessarily produce $C^1(\Gamma)$ variations δx_w, but only $L^2(\Gamma)$ and therefore we need to project them into $H^{(2n-1)/2}(\Gamma)$ for instance (an example of this is given in Fig. 6.3).

The fact that the gradient necessarily has less regularity than the parameterization is also easy to understand. Suppose that the cost function is a quadratic function of the parameterization: $J(x) = \|Ax - b\|^2$ with $x \in H^1(\Gamma)$, Ax and b in $L^2(\Gamma)$. The gradient $J'_x = 2A^T(Ax - b)$ belongs to $H^{-1}(\Gamma)$. Again, any parameterization variation using J'_x as descent direction will have less regularity than x: $\delta x = -\rho J'_x = -\rho(2(Ax - b)A) \in H^{-1}(\Omega)$. We therefore need to project (an engineer would say smooth) into $H^1(\Omega)$.

This situation is similar to what happens with CAD-free or level set parameters where a surface is represented by a large number (infinite) of independent points or functions evaluated at these points.

Now suppose the parameter belongs to a finite dimensional parameter space, as for instance with a polynomial definition of a surface. When we consider as parameter the coefficient of the polynomial, changes in the polynomial coefficients do not change the regularity as the new parameter will always belong to the same polynomial space. If the surface is parameterized by two (or several) polynomials, we need however to add regularity conditions for the junctions between the polynomials. We recover here the link introduced by the smoothing operator between parameter coefficients.

One way is to choose the projected variations $(\delta \tilde{x}_w)$ to be the solution of an elliptic system of degree $(2n-1)$. However, as we are using a P^1 discretization, a second-order elliptic system is sufficient even in 3D if we suppose the edges of the geometry (where the initial geometry is not C^1, for instance a trailing edge) as being constrained for the design. This means that they remain unchanged. Therefore we project the variations (δx_w) only into $H^2(\Gamma)$ even in 3D.

Hence, to avoid oscillations, we define the following local smoothing operator over the shape:

$$(I - \varepsilon \Delta)(\delta \tilde{x}_w) = \delta x_w, \qquad (6.1)$$

$$\delta \tilde{x}_w = \delta x_w = 0 \quad \text{on wedges},$$

where $\delta \tilde{x}_w$ is the smoothed shape variation for the shape nodes and δx_w is the variation given by the optimization tool. By local we mean that if the predicted shape is locally smooth, it remains unchanged during this step. The regions where the smoothing is applied are identified using a discontinuity-capturing operator.

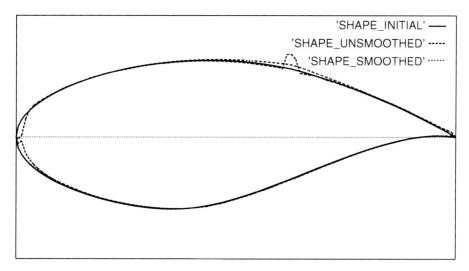

FIG. 6.3. Smoothed and nonsmoothed shapes. We can see that the gradient jumps through shocks and also produces a nonsmooth shape in leading edge regions. This is the result of the first iteration of the optimization. If we continue, the shape becomes more and more irregular.

Furthermore, the linear system is solved iteratively and ε is set to zero during the Jacobi iterations (in practice one performs a few iterations and stops when the residual is reduced by a few orders of magnitude which permits access to different levels of regularity) if

$$\frac{\delta_{ij}(\delta x_w)}{(\delta x_w)_T} < 10^{-3}, \tag{6.2}$$

where $\delta_{ij}(\delta x_w)$ is the difference between the variations of the two nodes of each segment of a surface triangle (nodes of a boundary edge in 2D) and $(\delta x_w)_T$ the mean variation on this triangle (edge in 2D).

This operation can also be seen as a modification of the scalar product for the Hilbert space in which optimization is performed and might therefore also have a preconditioning effect in the sense that it propagates localized high-frequency information where they are not seen at first hand. Hence, the impact of the smoothing in a descent algorithm:

$$J^{n+1} \leq J^n + (J'^n_x, \delta x)_0, \quad \delta x = -\rho J'^n_x,$$
$$J^{n+1} \leq J^n - \rho(J'^n_x, J'^n_x)_0 \leq J^n,$$

is located in the scalar product M (based on the projection operator for the gradient into the parameterization space) such that:

$$J^{n+1} \leq J^n - \rho(J'^n_x, J'^n_x)_M \leq J^n - \rho(J'^n_x, J'^n_x)_0.$$

Rotation, line search, and descent direction In addition to smoothing and preconditioning, the smoothing operator can be used for sensitivity evaluation (see Fig. 6.1). Suppose one has an approximate sensitivity and a rotation operator which operates over the unit sphere in the control space around a given control point. Our smoothing operator above can be such an operator if one projects the smoothed direction on the unit sphere (see Fig. 6.1). Consider the following smoothing-rotation operator over the unit sphere:

$$M(d,p)\tilde{d} = d.$$

M depends on d as the smoothing is only applied if necessary; p is the parameter defining the level of smoothing. Finding the right descent direction $\tilde{d}$ becomes a one-dimensional search problem for p:

$$p_{\text{opt}} = \text{argmin}_p J(x^n - \rho M^{-1}(d,p)d).$$

This is very useful for large dimensional problems and it also removes the need for specific development for gradient calculation. Once $\tilde{d}$ is defined it is used in the minimization algorithms described in Chapter 7. The definition of d can be, for instance, based on the incomplete sensitivity concept.

6.2.5 Level set

The level set method is an established technique to represent moving interfaces. Immersed boundary, fictitious domain methods, as well as penalizing methods, are methods to impose boundary conditions on surfaces which are not unions of edges and faces of elements of the (nonbody fitted) computational mesh [5, 6, 20, 21, 12, 14, 9, 2, 3, 24].

A parametrization of a boundary Γ by the level set method is based on the zero-level curve of a function ψ:

$$\Gamma = \{x \in \Omega \ : \ \psi(x) = 0\}.$$

The function ψ could be the signed Euclidean distance to Γ:

$$\psi(x) = \pm \inf_{y \in \Gamma} |x - y|,$$

with the convention of a plus sign if $x \in \Omega$ and minus sign otherwise. Hence

$$\psi|_\Gamma = 0, \quad \psi|_{\mathbf{R}^d \setminus \Omega} < 0, \quad \psi_\Omega > 0. \tag{6.3}$$

The definition can be extended to open shapes by using $\Gamma^\pm$ instead of Ω.

When the boundary moves with velocity V, i.e. at the next pseudo-time step $\zeta + \delta\zeta$:

$$\Gamma = \{x \ : \ \psi(\zeta + \delta\zeta, x + V\delta\zeta) = 0\},$$

then [23, 1, 17]:

$$\frac{\partial \psi}{\partial \zeta} + V\nabla\psi = 0.$$

For a given shape given by (6.3), the normal to Γ is $n = \nabla\psi/|\nabla\psi|$ at $\psi = 0$, useful for Neumann boundary conditions.

A relaxed signed characteristic function of Ω and its set-complement is

$$\chi = \psi/(|\psi| + \varepsilon_{opt}(h)), \qquad (6.4)$$

where $\varepsilon_{opt}(h)$ is a relaxation function strictly positive tending to zero with the background mesh size h. It is defined solving minimization problems for a sampling in h:

$$\varepsilon_{opt}(h) = \mathrm{argmin}_{\varepsilon(h)>0} \|u_h(\chi(\psi_h(\varepsilon(h)))) - \Pi_h\, u_{ref}\|,$$

where Π_h is the restriction operator to mesh h and u_h the discrete state. The numerical results given below have been computed with $\varepsilon_{opt}(h) = ch$ for some constant $c > 0$ (see Fig. 6.4). Above, u_{ref} is a reference solution which can be either a solution obtained with a body fitted mesh or, when available, an analytical solution.

Once ψ_h is known, we take into account the boundary conditions for a generic state equation $F_h(u_h) = 0$ using the equation

$$F_h(u_h)\chi(\psi_h) + F_{\Gamma_h}(u_h)\delta_{\psi_h} = 0.$$

Here F_{Γ_h} is the extension of the boundary condition for F_h on Γ_h over the domain and δ_ψ is a relaxed Dirac measure which is constructed using $\chi(\psi_h)$ and whose support approximates the boundary. It is interesting to notice that, when required, domain deformations are specified through $\chi(\Gamma_h)$. Therefore, in fluid-structure calculations for instance there is no need for a Lagrangian or Arbitrary Lagrangian Eulerian (ALE) [7] implementation. Likewise the complexity of deforming ALE body fitted meshes is avoided.

Figure 6.5 shows a comparison with an analytical solution [18, 4] for the scattering of a single pole acoustic source from a circular cylinder. One sees that the implicit representation of the shape by (6.4) is effective. It minimizes the effect of imperfect descriptions of the boundary on a given discretization and also avoids geometrical implementation complications.

Let us illustrate this immersed boundary method on a model problem. Consider

$$u - \nu u_{xx} = 0, \quad \text{on }]0, \pi[\qquad (6.5)$$

with boundary conditions

$$u_x(0) = 10, \quad u(\pi) = 1.$$

This can obviously be solved on a discrete domain with boundaries at 0 and π. But, if one uses the level set parameterization (6.4) to account for the boundary at $x = 0$, one solves for

$$\chi(u^\varepsilon - \nu u^\varepsilon_{xx}) + (1-\chi)(u^\varepsilon_x - 10) = 0, \qquad (6.6)$$

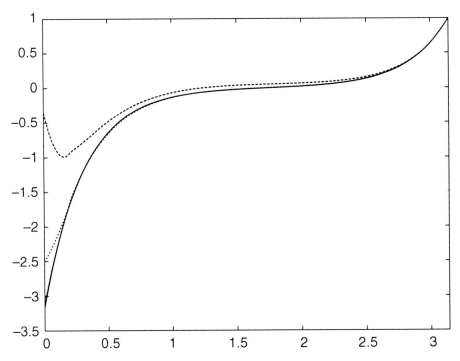

FIG. 6.4. Solution of model problem (6.5) (with $\nu = 0.1$) by immersed boundary method solving (6.6) for three different ε at given h. With $\varepsilon_{\text{opt}}(h) \sim 0.01h$ one recovers the target solution (the plain curve).

with χ defined here by

$$\chi = \frac{x}{x+\varepsilon},$$

which reduces the equation to

$$u^\varepsilon - \nu u^\varepsilon_{xx} + \frac{\varepsilon}{x}(u^\varepsilon_x - 10) = 0. \qquad (6.7)$$

One can see that this immersed boundary approach using a level set formulation is similar in solving a different equivalent continuous equation. Because the mesh does not fit the domain, this equation is solved on some slightly different domain such as $]\eta, \pi[$, where $0 \sim \eta < h \ll \pi$ if h is the mesh size. The nearly singular advection term convects into the domain $(x > 0)$ the information available at $x = 0$ with a decreasing advection intensity away zero. Solving (6.7) on a regular grid of size h will require upwinding stabilization for small[1] x. Adding a classical

---

[1] This is why oscillations sometimes appear when solving elliptic equations with an immersed boundary method if stabilization terms are not introduced when similar oscillations do not exist with body fitted approaches.

numerical viscosity term, the equivalent continuous equation for the discrete form reads:
$$u_h^\varepsilon - \left(\nu + \frac{h\varepsilon_h}{2x}\right)(u_h^\varepsilon)_{xx} + \frac{\varepsilon_h}{x}((u_h^\varepsilon)_x - 10) = 0. \tag{6.8}$$

To close the model one needs to specify $\varepsilon_h = \varepsilon_{\mathrm{opt}}(h)$ for the chosen h. A linear form in h for ε_h guarantees the consistency of the scheme. The coefficient c is fitted in order to minimize the error $\|u - u_h^\varepsilon\|$. The target solution u_{ref} comes here from a body fitted calculation (see Fig. 6.4).

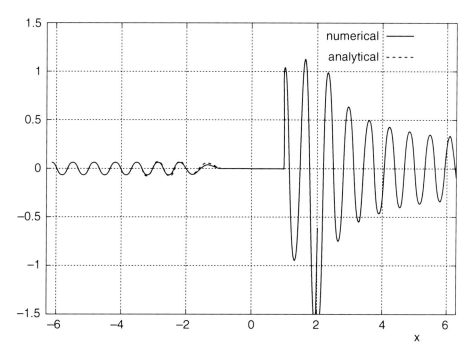

FIG. 6.5. Normalized instantaneous pressure: scattering of a monopole acoustic source placed at $(2,0)$ from a unit circular cylinder at a wave number of 10. The shape parameterization is based on formulation (6.4).

Figure 6.6 shows an example of flow calculation with the level set function for a complex geometry.

Regularity control for level set We describe how effectively this is handled in a gradient based shape optimization loop. Suppose an initial shape is given with corresponding level set parameterization ψ_0. Starting from this shape, a descent method based on direction d for ψ to minimize $J(\psi)$ can be seen as the discretization of (see also Chapter 7):

$$\psi_\zeta = -d(J(\psi)), \quad \psi_0 = \text{given}, \tag{6.9}$$

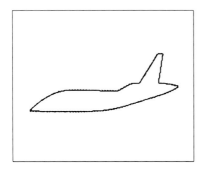 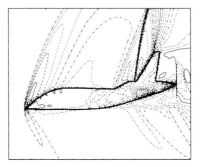

FIG. 6.6. Flow calculation with a relaxed normalized distance function to the shape.

by an explicit Euler scheme:

$$\psi_{n+1} - \psi_n = -\rho d_n, \quad \psi_0 = \text{given}. \tag{6.10}$$

Denoting by x the space coordinate, if one takes $d = -\nabla_\psi J |\nabla_x \psi|$ one recovers the Hamilton-Jacobi equation used for the motion of the level set function ψ [5, 20, 1]:

$$\psi_\zeta = -\nabla_\psi J |\nabla_x \psi| = -V \nabla_x \psi = 0, \quad V = \nabla_\psi J \cdot n, \tag{6.11}$$

where $n = \nabla_x \psi / |\nabla_x \psi|$. It is important to notice that both $\nabla_x \psi$ and $\nabla_\psi J$ are defined all over the domain and not only where $\psi = 0$ (the shape).

Hence, with this particular choice of d the algorithms for minimizing J and the motion of the level sets are the same. But, one should give attention to the regularity of the shape. Indeed, applying parameter variations proportional to the gradient of the functional will degrade the initial regularity of the level set parameterization. Therefore, the descent direction $d = -\nabla_\psi J |\nabla_x \psi|$ in (6.10) is regularized at each iteration using a regularization operator similar to (6.1) used for the CAD-free parameterization:

$$(I - \varepsilon \Delta) \tilde{d} = d, \tag{6.12}$$

$$\nabla_x \tilde{d}.n = 0 \quad \text{along external boundaries}.$$

This is solved with Jacobi iterations and one stops after the residual is reduced by one or two orders of magnitude. As in the CAD-free case, one can then access various shapes. Again, the regularization is over the whole domain and not only on the shape as for CAD-free where a Laplace-Beltrami operator was used.

Figure 6.7 shows an example of regularity control of the level set function in a shape optimization problem. Three iterations of the descent method have been applied for a drag reduction problem (see also Chapters 7 and 13). The different pictures show that regularization is necessary to avoid the dislocation of the shape. Figures 6.8 and 6.9 show how level set parameterization manages topology changes if regularity is monitored.

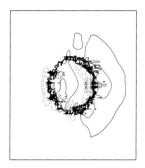 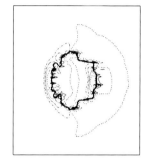 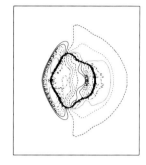

FIG. 6.7. Effect of the regularization operator on the level set function during optimization. Without regularization (left) and with increasing regularity requirement. Iso-lines show iso-$\chi(\psi)$ and iso-Mach contours.

 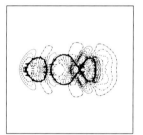 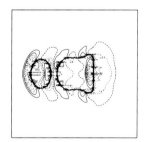

FIG. 6.8. Topology changes with level sets with regularity control.

6.3 Handling domain deformations

In cases where a fitted body approach is used (unlike with level sets) with the mesh fitting the shape, once $(\delta \tilde{x}_w)$ is known, we have to spread these variations over all the inner vertices. We give here a few algorithms with which to perform this task. We require the deformation method to preserve the positivity and orientation of the elements. In other words, the final mesh has to be conformal. Let us call D the deformation operator applied to the shape deformation (δx_w). We need the mesh deformation δx_m to satisfy $\delta x_m = \delta \tilde{x}_w$ on the shape, $\delta \tilde{x}_w$

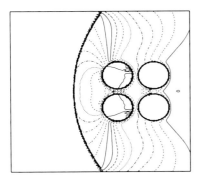 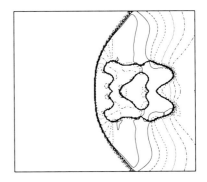

FIG. 6.9. Supersonic flows around complex bodies with level sets. Minimizing drag at zero-lift has led to topology changes and final symmetric configuration (without explicit requirement). A supersonic bug is more efficient than four UFOs.

being equal to δx_w on the shape and 0 elsewhere. This relation is therefore a boundary condition for D.

6.3.1 Explicit deformation

This is a simple way to explicitly prescribe the deformation to apply to each mesh node, knowing the shape deformation. The idea is to make the deformation for a node proportional to its distance to the shape. In other words, for an internal node i, we have:

$$(\delta x_m)_i = \frac{1}{\alpha_i} \sum_{k \in \Gamma_w} w_k \alpha_{ki} (\delta \tilde{x})_i, \qquad (6.13)$$

$$\delta \tilde{x_m} = \delta \tilde{x} \text{ on } \Gamma_w, \qquad (6.14)$$

where,
- δx_m is the variation of the mesh nodes;
- w_k is a weight for the contribution of each of the node k of the shape;
- $\alpha_{ki} = 1/|\tilde{x}_k - \tilde{x}_i|^\beta$ with β a positive arbitrary parameter;
- $\alpha_i = \sum_{k \in \Gamma_w} w_k \alpha_{ki}$ is the normalization parameter.

In 2D, w_k is equal to the sum of half of each segments size sharing the node k and in 3D equal to $1/3$ of the surface triangle sharing the node k. This algorithm has been widely used for the propagation of shape deformation [15, 13, 16] .

A justification for this algorithm can be found from the the following integral:

$$\frac{1}{\int_{\Gamma_w} \frac{d\gamma}{|x_m - x_w|^\beta}} \int_{\Gamma_w} \frac{\delta(x_w)}{|x_m - x_w|^\beta} d\gamma,$$

where $\int_{\Gamma_w} \delta(x_w)/|x_m - x_w|^\beta d\gamma$ is the convolution operator of the Green function $1/|x_m|^\beta$ with the Dirac function δ at point x_w of the shape Γ_w. The bigger

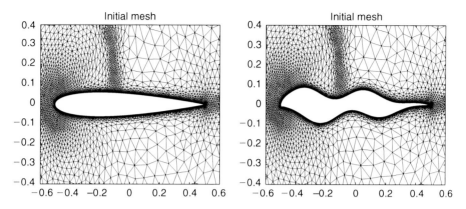

FIG. 6.10. Example of mesh deformation with the explicit deformation law (left: initial, right: deformed).

β is, the more the propagation is localized around the shape. For practical applications, $\beta = 4$ seems to be a good choice. This choice comes from experience and does not have a theoretical justification.

This algorithm is quite robust but expensive. Indeed, the complexity of the method is proportional to the number of shape discretization nodes times the number of mesh nodes. This is therefore too expensive to be used in 3D everywhere in the mesh. We usually use this approach close to the shape where the mesh is fine and continue the propagation with iterative methods based on the solution of elliptic systems (see Fig. 6.10).

6.3.2 Adding an elliptic system

Another possible choice is to solve an elliptic equation of the form of (6.1). Here, the different components are not coupled together:

$$(I - \varepsilon \Delta)\delta x_m = 0,$$
$$\frac{\partial \delta x_m}{\partial n} = 0 \quad \text{on slip boundaries,} \quad \delta x_m = \overline{\delta x_m} \quad \text{on walls.} \qquad (6.15)$$

None of the previous algorithms guarantee the conformity of the elements. To make this so, we need to introduce constraints on the deformations of the mesh nodes or for instance to make the viscosity ε of the previous elliptic system proportional to the inverse of the local mesh size. This is important so as to avoid mesh degeneration. The principle remains the same, however. We usually combine these two algorithms for close and far mesh fields.
  ICICI

6.3.3 Transpiration boundary condition

This is a classical way to account for small shape deformations without performing mesh deformations, by deriving a more complex boundary condition called an equivalent injection boundary condition, to be applied on the original shape. Denoting, $\vec{x}_w^1$ the original shape, $\vec{x}_w^2$ the shape after deformation, $\vec{u}$ being the flow velocity and $\vec{n}$ the unit normal to the shape, the slipping boundary condition can be expressed either in a fixed frame or in one attached and moving with the shape.

Denote by $\vec{n}_1$ and $\vec{n}_2$ the unit normal on each shape (in the fixed and moving frames). In frame 2, the slipping boundary condition reads: $\vec{u}_2.\vec{n}_2 = 0$ and in frame 1:

$$\vec{u}_2.\vec{n}_2 = \vec{V}.\vec{n}_2 = \frac{\vec{x}_w^2 - \vec{x}_w^1}{\delta t}.\vec{n}_2.$$

$\vec{V}$ is therefore the speed of the shape in frame 1. Now, if we suppose that the variations of the geometrical quantities dominate the physical ones (i.e. $|\delta x| \sim |\delta \vec{n}| \ll |\delta u|$):

$$\vec{V}.\vec{n}_2 = \vec{u}_2.\vec{n}_2 \sim \vec{u}_1.\vec{n}_1 + \vec{u}_1.(\vec{n}_2 - \vec{n}_1).$$

This defines an implicit relation for $\vec{u}_1.\vec{n}_1$ which is a new injection boundary condition for u in frame 1 relating the state at time $t + \delta t$ (time step $n+1$) of the computation to time t (time step n):

$$\vec{u}_1(t+\delta t).\vec{n}_1 = -\vec{u}_1(t).(\vec{n}_2(t+\delta t) - \vec{n}_1) + \vec{V}(t+\delta t).\vec{n}_2(t+\delta t).$$

In the same way, an equivalent boundary condition can be derived for the tangential velocity: $\vec{u}_2.\vec{\tau}_2 = 0$ in frame 2 and $\vec{u}_2.\vec{\tau}_2 = \vec{V}.\vec{\tau}_2$ in the fixed frame. A similar argument as above leads to:

$$\vec{u}_1(t+\delta t).\vec{\tau}_1 = -\vec{u}_1(t).(\vec{\tau}_2(t+\delta t) - \vec{\tau}_1) + \vec{V}(t+\delta t).\vec{\tau}_2(t+\delta t).$$

In these expressions, $\vec{\tau}_1, \vec{\tau}_2$ denote the local unit tangent vectors built from $\vec{n}_1, \vec{n}_2$. These tangent vectors are taken parallel to the local velocity in 3D.

With wall laws, the above expressions become $\vec{u}_2.\vec{\tau}_2 = u_\tau f(y^+)$ in frame 2 and $\vec{u}_2.\vec{\tau}_2 = u_\tau f(y^+) + \vec{V}.\vec{\tau}_2$ in the fixed frame, where u_τ denotes the friction velocity and $f(y^+)$ the wall function chosen which has to be adapted to the moving domain context. In particular, we need to introduce the shape velocity in the implicit relation which defines u_τ numerically. Hence, we have

$$\vec{u}_1(t+\delta t).\vec{\tau}_1 = u_\tau f(y^+) - \vec{u}_1(t).(\vec{\tau}_2(t+\delta t) - \vec{\tau}_1) + \vec{V}(t+\delta t).\vec{\tau}_2(t+\delta t).$$

These relations give satisfactory results when the shape curvature and the amount of the deformation are not high. They are therefore suitable for steady or unsteady shape optimization and control problems involving small shape deformations, as regions of large curvature are usually left unchanged due to constraints.

The previous approach is valid for both steady and unsteady configurations. However, in steady shape optimization we can proceed to some simplification. In particular, the terms involving the shape speed can be dropped and the geometry

is no longer time-dependent. The previous relations are therefore used in the following manner:
$$\vec{u}_1(t+\delta t).\vec{n}_1 = -\vec{u}_1(t).(\vec{n}_2 - \vec{n}_1),$$
$$\vec{u}_1(t+\delta t).\vec{\tau}_1 = u_\tau f(y^+) - \vec{u}_1(t).(\vec{\tau}_2 - \vec{\tau}_1).$$

As stated earlier, for unsteady configurations, when the amount of shape deformation is small, a good choice might be to prescribe an equivalent injection boundary condition. However, in multi-disciplinary applications (for instance in aeroelasticity) the amount of shape deformation due the coupling might be large and equivalent injection boundary conditions no longer sufficient. This is why for such applications, an Arbitrary Lagrangian Eulerian (ALE) formulation should be preferred to produce mesh deformation as for steady configurations and to take into account the mesh node velocity in the flow equations [7, 8, 19, 22].

6.3.4 Geometrical constraints

Geometrical constraints are of different types. We consider some of them:

- Defining two limiting surfaces (curves in 2D), shape variations are allowed between them.
- For shapes such as blade, wing or airfoil, the second constraint can be, for instance, that for the original plan-form remains unchanged.
- We can also require, for instance, that some parts of the geometry, such as the leading and trailing edges for wings, remain untouched.
- The previous constraints involve the parameterization points individually. We can introduce two global constraints for volume and thickness. We can require the volume and the maximum thickness for any section of the shape remain unchanged.

There are several ways to take into account equality and inequality constraints. The easiest way to treat the first, second, and third constraints is by projection:
$$\delta \tilde{x}_w = \Pi(\delta x_w).$$

The last constraint can be introduced in the optimization problem by penalty in the cost function ($J(x_w)$ being the cost function and $\alpha > 0$):
$$J(\tilde{x}_w) = J(x_w) + \alpha |V - V_0|.$$

The constraint on the thickness is more difficult to take into account. To this end, we define a by-section (see Fig. 6.11) definition of the shape where the number of sections required is free and depends on the complexity of the geometry. Then, each node in the parameterization is associated to a section Σ, and for each section, we define the maximum thickness Δ as:
$$\Delta(x_w) = x_w \to \Sigma \to \Delta.$$

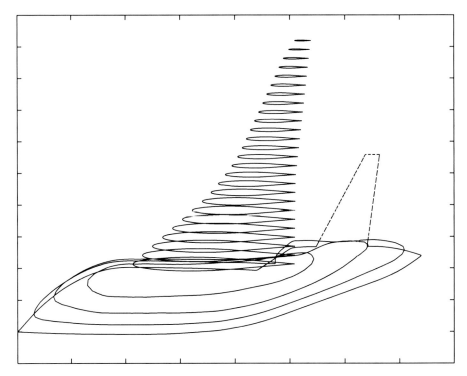

FIG. 6.11. By-section definition of an aircraft from its unstructured surface mesh.

Finally, we require the shape deformation to be in the kernel of the gradient of Δ, (i.e. the maximum thickness to have a local minima):

$$\nabla_{x_w}(\Delta(x_w)).\delta \tilde{x}_w = 0,$$

and therefore that $\Delta x_w = \Delta(x_w + \delta \tilde{x}_w)$. This means that we need a projection operator over $\mathrm{Ker}(\nabla_{x_w}(\Delta(x_w)).)$. We can also introduce this constraint in the cost function by penalty:

$$J(\tilde{x}_w) = J(x_w) + \alpha |V - V_0| + \beta |\Delta - \Delta_0|.$$

Given lift constraint Most aerodynamic shape optimizations are required to be at a given lift. It is interesting to observe that in a cruise situation (far from stall), the lift is linear with respect to the angle of incidence. This therefore gives another way of enforcing the given lift constraint. Indeed, during optimization the incidence follows the equation:

$$\theta^{n+1} = \theta^n - \alpha(C_l^n - C_l^{\mathrm{target}}), \quad 0 < \alpha < 0.5$$

where n is the optimization iteration.

In practice, especially when using incomplete state and sensitivity evaluations, it is not suitable to change either the inflow (angle of attack) or the shape incidence. Indeed, the first solution would require a long simulation for the flow solution to adapt to the new incidence and the second case would imply extra mesh deformation by rotating the whole shape. We have to keep in mind that a new incidence has to be realized at each optimization iteration. The best way to enforce the given lift constraint is to rotate the obtained gradient following the required incidence.

To avoid violation or saturation of inequality constraints interior point algorithms can be used instead of penalty formulation. These are based on modified BFGS problems to satisfy inequality constraints at each step of the optimization [11].

6.4 Mesh adaption

The following algorithms show how to couple shape optimization and mesh adaptation. To simplify the notation, the adaptive optimization algorithm is given with only one optimization after each adaptation, but several iterations of optimization can be made on the same mesh. The algorithms are presented with the simplest steepest descent method with projection. We will see that the definition of the projection operator is not an easy task.

At step i of adaptation, we denote the mesh, the solution, the metric and the cost $J(x_i, U(x_i))$ by $\mathcal{H}_i$, $\mathcal{S}_i$, $\mathcal{M}_i$ and $J(x_i)$.

Algorithm A1

$\mathcal{H}_0, \mathcal{S}_0$, given,

Adaptation loop: Do $\quad i = 0, ..., iadapt$

define the control space (wall nodes): $\mathcal{H}_i \rightarrow x_i$,

compute the gradient: $(x_i, \mathcal{H}_i, \mathcal{S}_i) \rightarrow \dfrac{dJ(x_i)}{dx_i}$,

define the deformation: $\tilde{x}_i = P_i \left(x_i - \lambda \dfrac{dJ(x_i)}{dx_i} \right)$,

deform the mesh: $(\tilde{x}_i, \mathcal{H}_i) \rightarrow \tilde{\mathcal{H}}_i$,

update the solution over the deformed mesh: $(\tilde{\mathcal{H}}_i, \mathcal{S}_i) \rightarrow \tilde{\mathcal{S}}_i$,

compute the metric: $(\tilde{\mathcal{H}}_i, \tilde{\mathcal{S}}_i) \rightarrow \mathcal{M}_i$,

generate the new mesh using this metric: $(\tilde{\mathcal{H}}_i, \mathcal{M}_i) \rightarrow \mathcal{H}_{i+1}$,

interpolate the previous solution over the new mesh:

$(\tilde{\mathcal{H}}_i, \tilde{\mathcal{S}}_i, \mathcal{H}_{i+1}) \rightarrow \overline{\mathcal{S}}_{i+1}$, compute the new solution over this mesh: $(\mathcal{H}_{i+1}, \overline{S}_{i+1}) \to S_{i+1}$,

End do

In the previous algorithm, the projection operator P_i changes after each adaptation as the control space changes. Therefore, the convergence issue is not clear. In the same way, adaptation to a more sophisticated pseudo-unsteady system based algorithm involving more than two shape parameterizations seems difficult. To avoid this problem, we define an exhaustive control space X suitable for a good description of any shape deformation. This can be associated to the first mesh used in the adaptation loop. In addition, this mesh has just to be refined in the vicinity of the wall (and can be coarse elsewhere). An example of this is given in Fig. 6.12. We will see that we can also have just a surface mesh.

We then use the following algorithm:

Algorithm A2

$\mathcal{H}_0, S_0, X$ given,

Adaptation loop: Do $i = 0, ..., iadapt$

identify the wall nodes: $\mathcal{H}_i \to x_i$,

compute the gradient: $(x_i, \mathcal{H}_i, S_i) \to \dfrac{dJ(x_i)}{dx_i}$,

interpolate the gradient $\dfrac{dJ}{dX} = \Pi_i \left(\dfrac{dJ(x_i)}{dx_i} \right)$,

define the deformation: $\tilde{X} = P\left(X - \lambda \dfrac{dJ}{dX} \right)$,

interpolate back the deformation $\tilde{x}_i = \Pi_i^{-1}(\tilde{X})$,

the rest as in A1:

deform the mesh: $(\tilde{x}_i, \mathcal{H}_i) \to \tilde{\mathcal{H}}_i$,

update the solution over the deformed mesh: $(\tilde{\mathcal{H}}_i, S_i) \to \tilde{S}_i$,

compute the metric: $(\tilde{\mathcal{H}}_i, \tilde{S}_i) \to \mathcal{M}_i$,

generate the new mesh using this metric: $(\tilde{\mathcal{H}}_i, \mathcal{M}_i) \to \mathcal{H}_{i+1}$,

interpolate the previous solution over the new mesh:

$(\tilde{\mathcal{H}}_i, \tilde{S}_i, \mathcal{H}_{i+1}) \to \overline{S}_{i+1}$,

compute the new solution over this mesh: $(\mathcal{H}_{i+1}, \overline{S}_{i+1}) \to S_{i+1}$,

End do

Here the projection P is defined once and for all, but the interpolation operator Π_i has to be redefined at each adaptation. However, we can follow the convergence of $|dJ/dX|$ as X is defined once and for all.

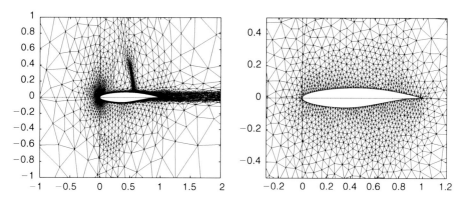

FIG. 6.12. The initial coarse mesh in the adaptation loop which is also used as background mesh to take into account the shape deformation. This is because we only need the shape, but we do not want to introduce an extra data structure. The exhaustive control space X is defined on this mesh at the beginning of the optimization loop. On the left, we have an intermediate adapted mesh for a transonic turbulent drag reduction problem over a RAE 2822 profile.

Another interesting feature of algorithm A2 is that it allows adaptation within the optimization loop. Indeed, as we said, in A1 we made the adaptation outside the optimization loop to avoid differentiating the mesh generator. Another way to avoid this is to use our approximate gradient approach, where we need only information over the shape. Therefore we can avoid taking into account dependencies created by the mesh generation tool inside the domain. However, we still need to keep the exhaustive control space X to be able to analyze the convergence of the algorithm. Hence, the last algorithm coupling all these ingredients is given by:

Algorithm A3

$\mathcal{H}_0, \mathcal{S}_0, X$ given,

Optimization loop: DO $\quad i = 0, ..., ioptim$

identify the wall nodes: $\mathcal{H}_i \to x_i$,

compute the approximate gradient: $(x_i, \partial \mathcal{H}_i, \partial \mathcal{S}_i) \to \dfrac{d\tilde{J}(x_i)}{dx_i}$,

interpolate the gradient: $\dfrac{d\tilde{J}}{dX} = \Pi_i \left(\dfrac{d\tilde{J}(x_i)}{dx_i} \right)$,

define the deformation: $\tilde{X} = P \left(X - \lambda \dfrac{d\tilde{J}}{dX} \right)$, interpolate back the deformation $\tilde{x}_i = \Pi_i^{-1}(\tilde{X})$,

deforme the mesh: $(\tilde{x}_i, \mathcal{H}_i) \to \tilde{\mathcal{H}}_i$,

update the solution over the deformed mesh: $(\tilde{\mathcal{H}}_i, \mathcal{S}_i) \to \tilde{\mathcal{S}}_i$,

Adaptation loop: DO $j = 0, ..., iadapt$

compute the metric: $(\tilde{\mathcal{H}}_j, \tilde{\mathcal{S}}_j) \to \mathcal{M}_j$,

generate the new mesh using this metric: $(\tilde{\mathcal{H}}_j, \mathcal{M}_j) \to \mathcal{H}_{j+1}$,

interpolate the previous solution over the new mesh:

$$(\tilde{\mathcal{H}}_j, \tilde{\mathcal{S}}_j, \mathcal{H}_{j+1}) \to \overline{\mathcal{S}}_{j+1},$$

compute the new solution over this mesh: $(\mathcal{H}_{j+1}, \overline{\mathcal{S}}_{j+1}) \to \mathcal{S}_{j+1}$,

End do adaptation

End do optimization

Where $\partial \mathcal{H}_i$ means mesh boundary informations and $\partial \mathcal{S}_i$) means the solution restricted to $\partial \mathcal{H}_i$.

6.5 Fluide-structure coupling

Let us consider a multidisciplinary fluid/structure problem where the state equations gather a fluid model in one side and on the other side an elastic model. We suppose the coupled problem has a stationary solution.

Denoting by $x(t)$ the CAD-free shape parameterization based on the fluid mesh, this can be summarized, for instance, through:

$$\frac{\partial U(x(t))}{\partial t} + \nabla_x.(F(x, U(x), \nabla_x U(x)) - S(U(x)) = 0 \quad \text{in } \Omega_f, \tag{6.16}$$

where U is the vector of conservative variables (i.e. $U = (\rho, \rho\vec{u}, \rho(C_v T + \frac{1}{2}|\vec{u}|^2))^t$, $\rho k, \rho\varepsilon)$, F represents the advective and viscous operators and S the right-hand side in the $k - \varepsilon$ turbulence model as described in Chapter 3. Ω_f denotes the fluid domain of resolution. This system has six equations in 2D (seven in 3D) and the system is closed using the equation of state $p = p(\rho, T)$.

To predict the behavior of the structure under aerodynamic forces, we consider the following time-dependent elastic model for the shape parameterization from the structural side $X(t)$:

$$M\ddot{X} + D\dot{X} + KX = F(U(x(t))|_\Gamma) \quad \text{in } \Omega_s, \tag{6.17}$$

where Ω_s denotes the structure domain of resolution and where M, D, and K are the mass, damping, and stiffness operators describing the structure characteristics. F contains the external forces acting on the structure and in particular the aerodynamic forces defined over $\Gamma = \partial\Omega_s \cap \partial\Omega_f$.

For instance, X represents nodes coordinates in a finite element mesh for the structure. Using this mesh, let us define for the structure a CAD-free parameterization of the shape: $X|_{\partial\Omega_s}(t)$. This is usually different from the one used for the shape seen from the fluid side. This is because the mesh used to solve (6.16) is different from the one used for (6.17). We have therefore to define transfer operators ($T_{s/f}$ and $T_{f/s}$) from/to $X(t)$ to/from $x(t)$. We need to express how to transfer structural deformation from X to x and inversely aerodynamic forces from x to X. Below, we choose x as control parameter; we also need therefore to transfer variations in x in X. Transfer operators are to enforce the action-reaction principle and energy conservation [7, 8, 19, 22].

Consider a minimization problem for a functional $J(x(t))$. To close the problem we need an equation for $x(t)$. For simplicity, consider the following pseudo-unsteady system representing a descent algorithm considered as an equation for the shape parameterization. Here, the time is fictitious and will play the role of the descent parameter:

$$\dot{x} = \Pi(-\nabla_x J),$$

where Π is the projection operator over the admissible space. Suppose (6.5) is discretized using a backward Euler scheme (denote by δx^n the shape deformation at step n):

$$\delta x^n = \Pi(-\lambda \nabla_{x^n} J^n),$$

where λ denotes the descent step size. The initial shape (x^0, X^0) given, the algorithm is as follows:

Do
- compute the gradient: $\nabla_{x^n} J^n$,
- define the new admissible shape deformation by (6.5),
- add to compatible structural deformation,

$$\delta x^n \leftarrow \delta x^n + T_{s/f} \delta X^n$$

- deform the fluid mesh: m_f^{n+1}
- enforce compatible structural deformation $\delta X^n = T_{f/s} \delta x^n$ such that

$$\int_f \sigma_f^n . n_f^{n+1} = \int_s \sigma_s^n . n_s^{n+1},$$

$$\int_f \dot{m}_f^{n+1} \sigma_f^n = \int_s \dot{m}_s^{n+1} \sigma_s^n,$$

$$\dot{m}_f^{n+1} = \frac{m_f^{n+1} - m_f^n}{\lambda},$$

- advance in time by λ the state equations: (U^{n+1}, X^{n+1})
- If $J^{n+1} <$ TOL, stop.

Done

Where, subscript f (resp. s) denotes fluid (resp. structural) quantities. $\sigma_{f/s}$ are the constraint tensors on the fluid and structure sides, respectively.

References

[1] Allaire, G. Jouve, F. and Toader, AM. (2002). A level-set method for shape optimization. *C. R. Acad. Sci. Paris*, Serie I, **334**, 1125-1130.

[2] Anderson, DM. McFadden, GB. and Wheeler, A. (1999). Diffuse-interface methods in fluid mechanics, *Annu. Rev. Fluid Mech.*, **30**, 139-165.

[3] Angot, P. Bruneau, CH. and Frabrie, P. (2001). A penalization method to take into account obstacles in viscous flows, *Numerische Mathematik*, **81**, 497-520.

[4] Arina, R. and Mohammadi, B. (2008). An immersed boundary method for aeroacoustic problems, *AIAA*, **08-3003**.

[5] Dervieux, A. and Thomasset, F. (1979). A finite element method for the simulation of Rayleigh-Taylor instability, *Lecture Notes in Mathematics*, Springer, Berlin, **771**, 145-159.

[6] Dervieux, A. and Thomasset, F. (1981). Multifuid incompressible fows by a finite element method , *Lecture Notes in Physics*, Springer, Berlin, **11**, 158-163.

[7] Donea, J. (1982). An ALE finite element method for transient fluid-structure interactions, *Comp. Meth. App. Mech. Eng.*, **33**, 689-723.

[8] Farhat, C. and Lesoinne, M. (1996). On the accuracy, stability and performance of the solution of three-dimensional nonlinear transient aeroelastic problems by partitioned procedures, *American Institute of Aeronautics and Astronautics*, **96-1388**.

[9] Glowinski, R. Pan, TW. and Periaux, J. (1998). A fictitious domain method for external incompressible viscous flows modeled by Navier-Stokes equations, *Comput. Meth. Appl. Mech. Eng.*, **112**, 133-148.

[10] Hicks, R. M. and Henne, P. A. (1978). Wing design by numerical optimization, *J. Aircraft*, **15**, 407-412.

[11] Herskovits, J.. (1986). A two-stage feasible directions algorithm for nonlinear constrained optimization , *Math. Program.*, **36**, 19-38.

[12] Kim, J. Kim, D. and Choi, H. (2005). An immersed-boundary finite-volume method for simulations of flow in complex geometries, *J. Comput. Phys.*, **171**, 132-150.

[13] Laporte, E. (1999). Shape optimization for unsteady configurations, PhD thesis, Ecole Polytechnique.

[14] Leveque, RJ. Li, Z. (1995).The immersed interface method for elliptic equations with discontinuous coefficients and singular sources, *SIAM J. Num. Anal.*, **31**, 1001-1025.

[15] Marrocco, A. and Pironneau, O. (1987). Optimum design with Lagrangian finite element, *Comp. Meth. Appl. Mech . Eng.*, **15(3)**, 512545.

[16] Medic, G. Mohammadi, B. Petruzzelli, N. and Stanciu, M. (1999). 3D optimal shape design for complex flows: application to turbomachinery, *American Institute of Aeronautics and Astronautics*, **99-0833**.

[17] Mohammadi, B. (2007).Global optimization, level set dynamics, incomplete sensitivity and regularity control, *IJCFD*, **21(2)**, 61-68.

[18] Morris, P.J. (1995). The scattering of sound from a spatially distributed axisymmetric cylindrical source by a circular cylinder, *J. Acoust. Soc. Am.*, **97(5)**, 2651-2656.

[19] Nkonga, B. and Guillard, H. (1994). Godunoc type method on non-structured meshes for three dimensional moving boundary problems, *Comp. Meth. Appl. Mech. Eng.*, **113**, 183-204.

[20] Osher, S. and Sethian, J. (1988). Fronts propagating with curvature-dependent speed: Algorithms based on Hamilton-Jacobi formulations, *J. Comp. Phys.*, **79(1)**, 12-49.

[21] Peskin, Ch. (1998). The fluid dynamics of heart valves: experimental, theoretical and computational methods, *Ann. Rev. Fluid Mech.*, **14**, 235-259.

[22] Piperno, S. Farhat, C. and Larrouturou, B. (1995). Partitioned procedures for the transient solution of coupled aeroelastic problems, *Comp. Meth. Appl. Mech. Eng.*, **124**, 97-101.

[23] Sethian, J.A. (1999). *Fast Marching Methods and Level Set Methods: Evolving Interfaces in Computational Geometry, Fluid Mechanics, Computer Vision and Materials Sciences*, Cambridge University Press.

[24] Verzicco, R. Fatica, M. Iaccarino, G. Moin, P. and Khalighi, B. (2001). Large eddy simulation of a road-vehicle with drag reduction devices, *AIAA J.* **40**, 2447-2455.

7
LOCAL AND GLOBAL OPTIMIZATION

7.1 Introduction

Most algorithms of optimization theory were invented for functionals of a finite number of variables (cradient, conjugate gradient, Newton and quasi-Newton methods). Most of them use optimal step sizes or Amijo rules with the objective to decrease the functional as much as possible at each step.

Our experience, and A. Jameson seems to think the same, is that optimal step size is not necessarily a good idea for shape optimization. It is better to think in terms of the dynamical system attached to the minimization process. This can be done for most of the algorithms used in practice namely first and second-order gradient-Newton methods, evolutionary algorithms and reduced-order modeling as we shall see. Thinking in these terms we intend to take advantage of the different available parameters of the methods and adapt the situation to shape optimization. Such an approach is heuristic and based on our numerical experience.

7.2 Dynamical systems

Consider the minimization of a functional $J(x) \geq 0$, $x \in \mathcal{O}_{ad}$. We suppose the problem admissible (i.e. there exist at least one solution x_m to the problem: $J(x_m) = J_m$). Most minimization algorithms [49] can be seen as discretizations of (see below for examples):

$$M(\zeta, x(\zeta))x_\zeta = d(x(\zeta)), \quad x(\zeta = 0) = x_0. \tag{7.1}$$

M is aimed to be positive definite and $M^{-1}d$ is built to be an admissible direction. In fact, we will see that even genetic algorithms can be cast in this form.

In the context of global optimization, it has been shown that algorithms issued from discretization of second-order systems [1,2] should be preferred over the discrete form of (7.1) as second-order dynamics permit us to escape from local minima:

$$\eta x_{\zeta\zeta} + M(\zeta, x(\zeta))x_\zeta = d(x(\zeta)), \quad x(\zeta = 0) = x_0, \quad x_\zeta(\zeta = 0) = x_0'. \tag{7.2}$$

7.2.1 Examples of local search algorithms

Obviously, with $\eta = 0, M = Id$ and $d = -\nabla J$, one recovers the steepest descent with vanishing step size:

$$x_\zeta = -\nabla J, \quad x(0) = x_0,$$

and with a uniform discretization one recovers a fixed step size:
$$x^{p+1} = x^p - \lambda^p \nabla J^p, \quad x^0 = x_0, \quad p \in \mathbb{N}, \quad \lambda^p = \lambda.$$

Recall that the step size used with gradient (and Newton) methods is uniform and defined for instance via an Armijo rule:

$$\text{given } (\sigma, \beta) \in (0,1)^2, \ \lambda_0 > 0, \quad \text{find } m \in \mathbb{N}, \quad \text{such that:}$$
$$J(x^p) + \sigma \beta^{m+1} \lambda_0 \nabla J^{pt} d^p \leq J(x^p + \beta^m \lambda_0 d^p)$$
$$\leq J(x^p) + \sigma \beta^m \lambda_0 \nabla J^{pt} d^p$$

Experience shows that the optimal descent step size is not always optimal for optimization problems if the geometry of the problem is unknown as one might be captured by the nearest local minima. Often, a constant descent step size should be preferred (see Figs. 7.1 and 7.2). Of course, by doing so we lose the theoretical convergence results.

With $M = \nabla^2 J$ and $\lambda^p \approx 1$ one recovers the Newton method:
$$x^{p+1} = x^p - \lambda^p (\nabla^2 J(x^p))^{-1} \nabla J^p, \quad x^0 = x_0, \quad p \in \mathbb{N}.$$

Quasi-Newton type methods can be expressed by introducing an approximation of the Hessian, for instance, an iterative formula such as the one due to Broyden, Fletcher, Goldfarb and Shanno (BFGS) where $(\nabla_{xx} J^p)$ is approximated by a symmetric positive definite matrix $\mathbf{H}^p$, constructed iteratively, starting from the identity matrix:

$$\mathbf{H}^{p+1} = \mathbf{H}^p + \frac{\gamma^p \gamma^{p^T}}{\gamma^{p^T} \delta^p} - \frac{\mathbf{H}^p \delta^p (\delta^p)^T (\mathbf{H}^p)^T}{(\delta^p)^T \mathbf{H}^p \delta^p}, \tag{7.3}$$

with
$$\gamma^p = \nabla J(x^{p+1}) - \nabla J(x^p), \quad \delta^p = x^{p+1} - x^p.$$

Less intuitively, one can also recover methods such as conjugate gradient which corresponds to the following discrete form:

$$x^{p+1} - 2x^p + x^{p-1} = -\lambda^p \|g^p\| \sum_{i=1}^{p} \frac{g^i}{\|g^i\|} + \lambda^{p-1} \|g^{p-1}\| \sum_{i=1}^{p-1} \frac{g^i}{\|g^i\|}, \tag{7.4}$$

where $g^i = \nabla_x J^i$. After simplification, one has:

$$x^{p+1} - 2x^p + x^{p-1} = -\lambda^p g^p + \|g^{p-1}\| \sum_{i=1}^{p-1} (\lambda^{p-1} - \lambda^p) \frac{g^i}{\|g^i\|}, \tag{7.5}$$

This is a second-order integro-differential discrete system which can be seen as a discrete form of:

$$x_{\zeta\zeta} = \frac{-\nabla J(T_2)}{\lambda_2} + \frac{(\lambda_1 - \lambda_2)}{\lambda_2^2} \|\nabla J(T_1)\| \int_0^{T_2} \frac{\nabla J(\tau)}{\|\nabla J(\tau)\|} d\tau,$$

where $\lambda_i = \text{argmin}_\lambda (J(T_i) - \lambda \nabla J(T_i))$, $i = 1, 2$ and with $T_1 < T_2$.

Note that if λ is fixed in (7.5) and if $||g^p|| = ||g^{p-1}|| = g$ we have:

$$x^{p+1} - 2x^p + x^{p-1} = -\lambda g,$$

which describes the motion of a heavy ball without friction and does not have a stationary solution. This shows the importance of an optimal step size in the conjugate gradient method. This situation corresponds to the motion of a ball over a curve defined by an absolute value function smoothed around zero.

In Chapter 6 we showed the application of some of these algorithms with incomplete sensitivity analysis to aerodynamic shape optimization.

7.3 Global optimization

The global solution of (7.1) can be thought of as finding $x(T)$ satisfying:

$$M(\zeta, x(\zeta))x_\zeta = d(x(\zeta)), \quad x(0) = x_0, \quad J(x(T)) = J_m, \quad \text{with } T < \infty. \quad (7.6)$$

This is an overdetermined boundary value problem and it tells us why one should not solve global optimization problems with methods which are discrete forms of Cauchy problems for first-order differential systems except if one can provide an initial condition in the attraction basin of the global optimum. Hence, in the context of global optimization the initial condition $(x(0) = x_0,)$ is usually misleading.

To remove the overdetermination, we consider second-order systems. Local minimization can be seen then as finding $x(T)$ solution of:

$$\eta x_{\zeta\zeta} + M(\zeta, x(\zeta))x_\zeta = d(x(\zeta)), \quad x(0) = x_0, \quad J'(x(T)) = 0. \quad (7.7)$$

If in addition J_m is known, the final condition in (7.7) can be replaced by $J(x(T)) = J_m$.[2] These second-order systems with implicit final condition can be solved using solution techniques for boundary value problems with free surface [3,43,23–25]. An analogy can be given with the problem of finding the interface between water and ice which is only implicitly known through the iso-value of zero temperature.

From a practical point of view, success in global optimization means being able to reach in finite time the vicinity (a ball of given radius) of the infimum starting from any point in the control space. In this case, the associated boundary value problem is of the form:

$$\begin{cases} \eta x_{\zeta\zeta} + M(\zeta, x(\zeta))x_\zeta(\zeta) = d(x(\zeta)), \\ x(0) = x_0, \quad x_\zeta(0) = x_{\zeta,0} \\ J(x(T_{x_0})) \approx J_m. \end{cases} \quad (7.8)$$

Now, system (7.8) is overdetermined by $x_{\zeta,0}$.

---
[2] Actually, the initial condition might be dropped considering only final conditions: $J(x(T)) = J_m$, $J'(x(T)) = 0$.

Global optimization

But, it can be proved [42] that:
If $J : R^n \to R$ is a C^2-function such that $\min_{R^n} J$ exists and is reached at $x_m \in R^n$, then for every $(x_0, \varepsilon) \in \mathbf{R}^n \times \mathbf{R}^+$ there exists $(\sigma, T) \in \mathbf{R}^n \times \mathbf{R}^+$ such that the solution of the following dynamical system:

$$\begin{cases} \eta x_{\zeta\zeta}(\zeta) + x_\zeta(\zeta) = -J'(x(\zeta)) \\ x(0) = x_0, \\ x_\zeta(0) = \sigma, \end{cases} \quad (7.9)$$

passes for $\zeta = T$ into the ball $B_\varepsilon(x_m)$.
Below, we shall use this idea to derive recursive global optimization algorithms.

7.3.1 Recursive minimization algorithm

As said above, global optimization of J has a solution if, for a given precision ε in the functional (see Table 7.3.1), one can build at least one trajectory $(x(t), 0 \leq t \leq T_\varepsilon)$ passing in finite time T_ε close enough to x_m. In what follows, these trajectories will be generated by first or second-order Cauchy problems starting from an initial value v. These will be denoted $(x_v(t), 0 \leq t \leq T_\varepsilon, v \in \mathcal{O}_{ad})$. Above, we saw that existing minimization methods build such trajectories in discrete form. This can be summarized as:

$$\forall\, \varepsilon > 0, \exists\, (v, T_\varepsilon) \in \mathcal{O}_{ad} \times [0, +\infty[\quad \text{such that} \quad J(x_v(T_\varepsilon)) - J_m \leq \varepsilon. \quad (7.10)$$

If J_m is unknown T_ε defines the maximum calculation complexity wanted. In these cases, setting $J_m = -\infty$, one retains the best solution obtained over $[0, T_\varepsilon]$. In other words, one solves:

$$\forall\, (\varepsilon, T_\varepsilon) \in \mathbf{R}^+ \times [0, +\infty[, \exists\, (v, \tau) \in \mathcal{O}_{ad} \times [0, T_\varepsilon] \quad \text{such that} \quad J(x_v(\tau)) - J_m \leq \varepsilon. \quad (7.11)$$

A recursive algorithm can be used to solve (7.10) or (7.11):
Given $\varepsilon > 0$, J_m and $0 \leq T_\varepsilon < \infty$, we minimize the functional $h_{\varepsilon, T_\varepsilon, J_m} : \mathcal{O}_{ad} \to \mathbf{R}^+$ below:

$$h_{\varepsilon, T, J_m}(v) = \min_{x_v(\tau) \in \mathcal{O}_{ad},\, \tau \in [0, T]} (J(x_v(\tau)) - J_m) \quad (7.12)$$

using one of the local search algorithms mentioned above.

Once this core minimization problem is defined, global minimization is seen as a nested minimization problem where one looks for v to provide the best initial value to generate the trajectory $(x_v(\tau), 0 \leq \tau \leq T_\varepsilon)$, i.e. passing closer to x_m. For instance, the following recursive algorithm proposes a solution to this problem (denote $h_{\varepsilon, T_\varepsilon, J_m}$ by h).

Consider the following algorithm $A_1(v_1, v_2)$:

- $(v_1, v_2) \in \mathcal{O}_{ad} \times \mathcal{O}_{ad}$ given
- Find $v \in \mathrm{argmin}_{w \in \mathcal{O}(v_1, v_2)} h(w)$ where $\mathcal{O}(v_1, v_2) = \{v_1 + t(v_2 - v_1), t \in \mathbf{R}\} \cap \mathcal{O}_{ad}$ using a line search method
- return v

TABLE 7.1. Complexity evolution (number of evaluations for the pair (J, J')) in global optimization for the Rastrigin function $J(x) = N + \sum_{i=1}^{N}(x_i^2 - \cos(18x_i))$, $x \in [-5,5]^N$ for two values of accuracy, ε, required. We consider control spaces with dimension up to 10^4.

	$N=10$	$N=100$	$N=1000$	$N=5000$	$N=10000$
$\varepsilon = 10^{-3}$	O(50)	O(100)	O(100)	O(200)	O(400)
$\varepsilon = 10^{-6}$	O(50)	O(100)	O(100)	O(250)	O(700)

The minimization algorithm in A_1 is defined by the user. It may fail. For instance, a secant method degenerates on plateaux and critical points.[3] In this case, we add an external layer to the algorithm A_1 minimizing $h'(v') = h(A_1(v', w'))$ with w' chosen randomly in $\mathcal{O}_{ad}$. This is the stochastic feature of the algorithm.

This leads to the following two-layer algorithm $A_2(v'_1, v'_2)$:

- $(v'_1, v'_2) \in \mathcal{O}_{ad} \times \mathcal{O}_{ad}$ given
- Find $v' \in \mathrm{argmin}_{w \in \mathcal{O}(v'_1, v'_2)} h'(w)$ where $\mathcal{O}(v'_1, v'_2) = \{v'_1 + t(v'_2 - v'_1), t \in \mathbf{R}\} \cap \mathcal{O}_{ad}$ using a line search method
- return v'

Again, the search algorithm in A_2 is user-defined. For the algorithm to be efficient A_1 and A_2 need to be low in complexity [4].

This construction can be pursued by building recursively:

$h^i(v_1^i) = h^{i-1}(A_{i-1}(v_1^i, v_2^i))$, with $h^1(v) = h(v)$ and $h^2(v) = h'(v)$ where $i = 1, 2, 3, ...$ denotes external layers.

Due to the stochastic feature mentioned, this is a semi-deterministic algorithm [43, 23–25, 22].

7.3.2 Coupling dynamical systems and distributed computing

Several dynamical systems can be coupled in order to increase their global search feature and also to take advantage of parallel computing possibilities.

Suppose one solves the pseudo-unsteady system (7.6) by the recursive algorithm above where the aim is to provide a suitable initial guess to the corresponding Cauchy problem (7.2). Now, starting from different positions in the admissible space, one couples the paths using cross-information involving a global gradient.

Consider, q individuals $(x^j, j = 1, ..., q)$ with their motions prescribed by q pseudo-unsteady systems:

---

[3] In large dimensional problems one would rather use the Broyden method which is the multidimensional analogue of the secant method $x^{p+1} = x^p + (H^{p+1})^{-1}(J'^{p+1} - J'^p)$ with H^{p+1} based on an approximation of the Hessian of the functional using the BFGS formula for instance.

[4] Another good candidate also not limited by the one-dimensional search feature of linesearch is using data interpolation and a response surface as mentioned in section 7.6.1 using available information collected at lower levels. Overall it is true that optimization algorithms do not use enough previously available information.

$$\eta x^j_{\zeta\zeta} + M x^j_\zeta = d^j + G^j, \quad x^j(\zeta=0) = x^j_0, \quad x_\zeta(\zeta=0) = 0. \tag{7.13}$$

G_j is a global gradient representing the interaction between individuals:

$$G^j = -\sum_{k=1, k\neq j}^{q} \frac{J^j - J^k}{||x^j - x^k||^2}(x^j - x^k), \quad \text{for} \quad j = 1, ..., q.$$

This repulsive force aims at forcing the search to take place in regions not already visited. The right-hand side in (7.13) is therefore a balance between local and global forces. To be efficient for distributed calculation the communications between individuals should be weak. One possibility to reduce the exchanges of (x^j, J^j) between individuals is to replace the current position (x^j, J^j) by (x^j_{opt}, J^j_{opt}) indicating the best element found by the recursive algorithm for individual j. The communication is done therefore once every time the core minimization algorithm is completed in (7.12).

The behavior of search methods based on first and second-order systems (7.13) with two individuals with constant coefficients is shown in Fig. 7.1 for the Griewank function:

$$J(x,y) = 1 - \cos(x)\cos(\frac{y}{\sqrt{2}}) + \frac{1}{50}((x - y/2)^2 + 7y^2/4), \quad \text{for } (x,y) \in]-10, 10[^2.$$

Second-order dynamics improve global minimum search by helping to escape from local minima (see Fig. 7.1). Coupling two individuals helps find the global minimum (see Fig. 7.3).

7.4 Multi-objective optimization

Engineering design often deals with multiple and often conflicting objective functions $\tilde{\mathbf{J}} = \{J_i(x), i = 1, ..., n\}$ [44, 11, 9, 10, 48, 52, 29]. A very simple example is the design of a wing with maximum volume and minimum surface. In Chapter 10 we will see an example of civil supersonic aircraft design with minimal drag and sonic boom requirements which are conflicting. More generally one usually wants to maximize the performance of a system while minimizing its cost. As mentioned in Chapter 5 it is unlikely that the optimum for the different functionals coincides. One rather looks for a trade-off between conflicting objectives.

One optimality concept is known as Pareto equilibrium [44] defined by non-dominated points: A point x is dominated by a point y if and only if $J_i(y) \leq J_i(x)$, with strict inequality for at least one of the objectives. Thus, a global Pareto equilibrium point is such that no improvement for all objectives can be achieved (i.e. by moving to any other feasible point).

The classical method for multi-objective optimization is the weighted sum method [9, 10, 48]:

$$J(x) = \Sigma \alpha_i J_i(x), \quad \Sigma \alpha_i = 1, \quad \alpha_i \in [0, 1].$$

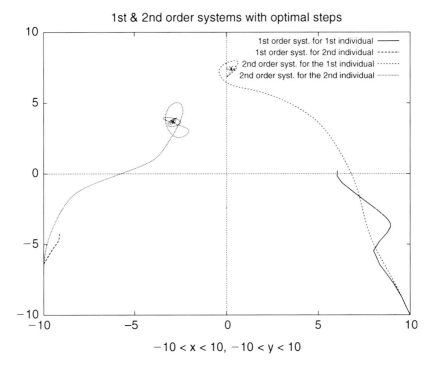

FIG. 7.1. Minimization for the Griewank function. The global minimum at $(0,0)$. Paths for the steepest descent and heavy ball methods starting from two different points.

A sampling in α_i enables us to reach various points in the Pareto front using a descent method for $J(x)$. This method is extensively used because one knows that it has a solution for convex functionals.

It is admitted that the weighted sum approach does not find Pareto solutions in nonconvex regions of the Pareto front [38]. A geometric interpretation is that when the tangent to the Pareto front is parallel to one of the axes the corresponding functional has a plateau which cannot be crossed by a descent method.

In case the front is nonconvex but the tangent is never parallel to one of the axes, one can still use a descent method for increasing power n in the functional [38]:

$$\tilde{J}(x) = \Sigma \alpha_i J_i^n(x), \quad \Sigma \alpha_i = 1, \quad \alpha_i \in [0,1].$$

One looks for the lowest n making the Pareto front convex because this approach has the disadvantage of degrading the conditioning of the optimization problem.

It is shown that alternative solutions, based on global optimization, exist [42]. This in particular shows that the failure of the weighted sum approach is due to

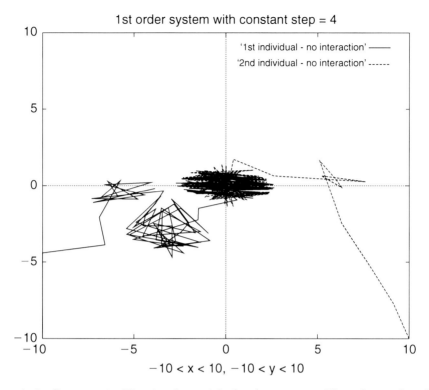

FIG. 7.2. Same as in Fig. 7.1 but with fixed step size. This shows that for nonconvex problems, a constant step size might overperform optimal step sizes.

a lack of a global minimization feature in descent methods (this remains valid with a Nash definition of the equilibrium).

We saw how to link any couple of points in the admissible space in finite time with some given accuracy by a trajectory solution of a second-order dynamical system such as (7.7). If the problem is admissible, a point in a Pareto front corresponds to a point in the admissible control space. It should be reached in finite time from any initial condition in the admissible space by a trajectory solution of our second-order system. This means that one can use a minimization method which has global search features to escape local minima in the case of nonconvex Pareto fronts.

To illustrate this let us consider the functionals $J_1 = x_1 + x_2$ and $J_2 = (1/x_1 + \|x\|^2) + \beta(\exp(-100(x_1-0.3)^2) + \exp(-100(x_1-0.6)^2))$ for $x \in [0.1, 1]^2$. Figure 7.4 shows Pareto front generation for $\beta = 0$ and 3. The first situation can be solved using a steepest descent method with optimal step size while the second case requires a global optimization method to reach the nonconvex region in the front. We have solved 10 minimization problems based on uniform sampling of

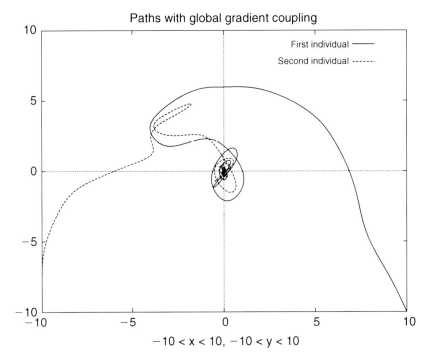

FIG. 7.3. Same minimization as in Fig. 7.1 but with the repulsive forces accounted for in each system.

$\alpha \in [0,1]$ in the weighted sum. We have then solved the minimization problems starting from the same point $(x(t=0) = (1,1)^t)$ in the admissible space using an algorithm based on discrete forms of second-order (7.9) dynamical systems with $J' = \alpha \nabla J_1 + (1-\alpha) \nabla J_2$ and with $\eta = 0$ and $\eta = 1$. For the second-order system the initial impulse is a variable as mentioned in (7.9). On sees that second-order dynamics is necessary to recover general Pareto fronts.

7.4.1 Data mining for multi-objective optimization

The trade-off between functionals can be represented by two-dimensional plots or three-dimensional surfaces if the optimization problem involves less than three objective functions. When the number of functionals and constraints is high one would need data mining techniques permitting a reduction in dimension in plots. One successful technique is the Self-Organizing Map (SOM) [30] which can be seen as a nonlinear extension of principal component analysis [20, 27] with two or three components.[5]

---

[5] Principal component analysis (PCA) produces an orthogonal linear coordinate transformation of the data optimum in a least squares sense. Projecting the data in this coordinate system, the variance decreases with the dimension. The first coordinate is the first PC.

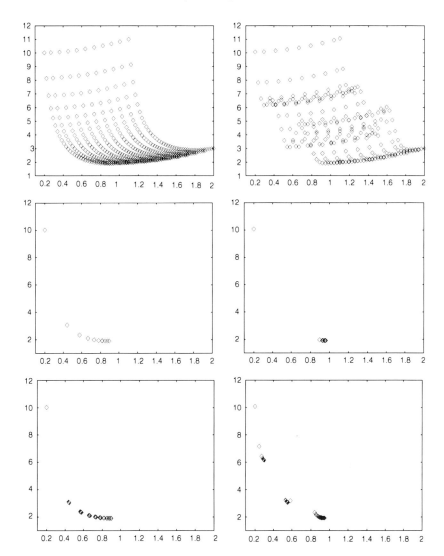

FIG. 7.4. Computation of Pareto fronts for (J_1, J_2). Top: one generates the Pareto front by sampling the control space; here $\beta = 0$ and 3. Middle and bottom: Solutions of minimization by the first (middle) and second-order (bottom) dynamical systems method for a uniform 10 point sampling of $\alpha \in [0, 1]$.

Suppose one has n functionals $\vec{J} = (j_1, ..., j_n)^t \in \mathcal{R}^n$ and the two first principal components or directions known.[6] Starting from a uniform sampling (model)

---

[6]If the PCA is not available one proceeds with a random initialization.

($\vec{m}_i, i = 1, ..., N$) of this two dimensional subspace, the SOM algorithm builds a map iteratively in two steps. At each iteration $k = 1, 2, ...$, for $\vec{J}$ from the available sample:

- find the best matching unit (BMU) between $\vec{m}_i^k$ ($\|.\|$ being the Euclidean distance in $\mathcal{R}^n$):

$$\|\vec{J} - \vec{m}_{\text{BMU}}^k\| \leq \|\vec{J} - \vec{m}_i^k\| \quad \forall i = 1, ..., N$$

- update locally the model using J:

$$\vec{m}_i^k \leftarrow \vec{m}_i^k + \alpha \exp(-s_k \|\vec{m}_{\text{BMU}}^k - \vec{m}_i^k\|) \left(\vec{J} - \vec{m}_i^k\right) \quad \forall i = 1, .., N.$$

s_k in the localization function is initiated at 0 and increases with k and $0 < \alpha < 1$. One chooses an a priori number of iterations and these iterations (regression) are repeated over the available sample until an equilibrium is reached in the map. The final map highly depends on the initialization. This map can be clustered using the Euclidean metric defining regions of closest $\vec{J}$. This clustering together with different colors for each j_i from the sample highlights different possible behavior as we see below for a shape optimization problem.

SOM has been employed successfully to identify the trade-off relations between objectives in aeronautics applications [26, 33, 45]. Figure 7.5 shows an example of such a map for a problem with three objective functionals (Drag C_d, wing weight, $-C_{p,\max}$) [34]. The data basis has been generated after a uniform sampling of the design space (wing-nacelle-pylon shape parameterized using NURBS). The results have been projected onto a two-dimensional map. A clustering of this map provides an estimation of the different behavior for the different shapes. After coloring the map with the different functionals, it can be seen that the lower right corner corresponds to low drag, low $-C_{p,\max}$ and wing weight. Shapes having their vector of functionals represented in this area are therefore suitable for all three functionals.

7.5 Link with genetic algorithms

We would like to show that genetic algorithms (GA) can be seen also as discrete forms of a system of first-order coupled ordinary differential equations with stochastic coefficients. Therefore, the boundary value problem formulation of global optimization can be applied to GA for a better definition of the search family.

Consider the minimization of a real functional $J(x)$, called the fitness function in the GA literature; as usual $x \in \Omega_{ad}$ means the optimization parameter and belongs to an admissible space Ω_{ad}, say of dimension $n_{\dim}$. Genetic algorithms approximate the global minimum (or maximum) of J through a stochastic process based on an analogy with the Darwinian evolution of species [19, 17, 16].

A first family, called population, $X^1 = \{(x)_l^1 \in \Omega_{ad}, l = 1, ..., N_p\}$ of N_p possible solutions in Ω_{ad} of the optimization problem, called individuals, is randomly generated in the search space Ω_{ad}. Starting from this population X^1 we

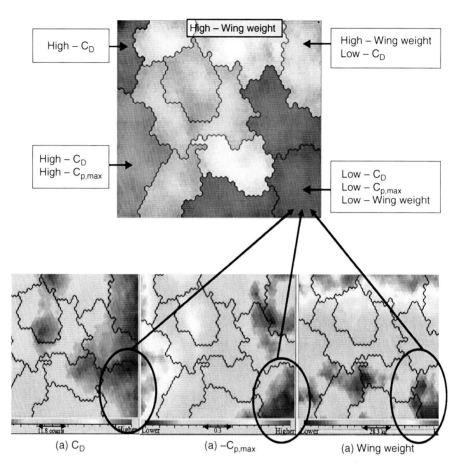

FIG. 7.5. Self-Organizing Map. (Courtesy of Prof. S. Obayashi, with permission)

build recursively N_{gen} new populations $X^i = \{(x)_l^i \in \Omega_{ad}, l = 1, ..., N_p\}$ with $i = 1, ..., N_{\text{gen}}$ through three stochastic steps where at each iteration i, to create a new population X^{i+1}, the following steps are taken:

Selection: We aim at removing the weakest individuals. Each individual x_l^i is ranked with respect to the fitness function J: the lower is J, the higher is the rank. Better individuals have higher chances to be chosen and become parents. This defines an intermediate population:

$$X^{i+1/3} = \mathcal{S}^i X^i, \tag{7.14}$$

where $\mathcal{S}^i$ is a binary (N_p, N_p) matrix with $\mathcal{S}^i_{j,j} = 1$ if the individual j is selected as parent and 0 otherwise.

Crossover: this step leads to a data exchange between parents and apparition of new individuals called children. We determine, with probability p_c if two consecutive parents in $X^{i+1/3}$ should exchange data or if they are copied directly into the intermediate population $X^{i+2/3}$:

$$X^{i+2/3} = \mathcal{C}^i X^{i+1/3}, \qquad (7.15)$$

where $\mathcal{C}^i$ is a real-valued (N_p, N_p) matrix. For each couple of consecutive lines $(2j-1, 2j)$ $(1 \leq j \leq N_p/2$, suppose N_p even), the coefficients of the $2j-1^{\text{th}}$ and $2j^{\text{th}}$ rows are given by:

$$C^i_{2j-1,2j-1} = \lambda_1, \quad C^i_{2j-1,2j} = 1 - \lambda_1, \quad C^i_{2j,2j-1} = \lambda_2, \quad C^i_{2j,2j} = 1 - \lambda_2,$$

where $\lambda_1 = \lambda_2 = 1$ if no crossover is applied on the selected parents and otherwise randomly chosen in $]0, 1[$ with a probability p_c.

Mutation: individuals are mutated with a given probability p_m. Introduce a random (N_p, N_{ndim}) perturbation matrix $\mathcal{E}^i$ with line j made of either an admissible random vector of size n_{dim} if a mutation is applied to individual j and otherwise a null vector. The mutation step is then:

$$X^{i+1} = X^{i+2/3} + \mathcal{E}^i.$$

These three steps give a particular discretization of a set of nonlinear first-order dynamic systems:

$$X^{i+1} - X^i = (\mathcal{C}^i \mathcal{S}^i - I)X^i + \mathcal{E}^i, \quad X^1 = \text{given}.$$

The scalar case is when $N_p \to 1$. Also, one can use the boundary value formulation (7.7) for global optimization to reduce the size of the family considered in GA [22]. It is indeed observed that if the family is initially well chosen GAs have a better performance.

Engineers like GAs because these algorithms do not require sensitivity computation, so the source code of the implementation is not needed; they can, in principle, perform global and multi-objective optimization and are easy to parallelize. Their drawbacks are the computing time for a given accuracy and possible degeneracy. It is also somewhat surprising that GAs have no inertial effects. One improvement could therefore be to add the missing second-order dynamics. Concerning degeneracy issues and revisiting of already visited minima, the cure seems to be on coupling with learning techniques and parametric optimizations. Also, as full convergence is difficult to achieve with GAs, it is recommended to complete GA iterations by descent methods (hybridization). This is especially useful when the functional is flat around the infimum [22, 14] as for the example shown in Section 5.4.

Table 7.5 shows the solution to Rastrigin functional minimization for three increasing control space dimensions (10, 100, 1000) using the semi-deterministic algorithm above (results from Fig. 7.3.1), a genetic algorithm and a hybridization of both [22].

Overall one can see that there is not a simple unique approach to optimization. Rather one should try to combine the different approaches.

TABLE 7.2. Rastringin functional. Total number of functional evaluations to reach the infimum with an accuracy of 10^{-6}.

	SDA	GA	Hybrid
N=10	O(50)	6000	2600
N=100	O(100)	O(1e5)	O(1e4)
N=1000	O(100)	O(1e7)	O(1e5)

7.6 Reduced-order modeling and learning

Due to prohibitive calculation cost of direct simulation loops and also the presence of noise and uncertainties it is useful to avoid calls to the direct loop. One efficient approach consists in using reduced-order models. This receives different denominations following the field of research it is issued from: learning in neural networks [37], fitting with least squares in experimental plans [21, 35] or higher degrees polynomials [47], reduced-order modeling with proper orthogonal decomposition [50] or other low-complexity modeling. We will see examples of such for sensitivity analysis using the incomplete sensitivity concept in Chapter 8. Below we give a general view of what we call reduced-order modeling.

Consider the calculation of a state variable by solving:

$$F(V(p)) = 0.$$

$V(p)$ is function of independent variables p. Our aim is to define a suitable search space for the solution $V(p)$ instead of considering a general function space.

This former approach is what one does in finite element methods, for instance, where one looks for a solution V_h expressed in some finite-dimensional subspace $S_N(\{W_i, i = 1, ..., N\})$:

$$V_h = \sum_{i=1}^{N} v_i^{\text{opt}} W_i.$$

S_N is generated by the functional basis chosen $\{W_i, i = 1, ..., N\}$. For instance, one can consider W_i as a polynomial of degree one ($W_i \in P^1$) on each element of a discrete domain called the mesh. v_i^{opt} denotes the values of the solution on the nodes of the mesh solution of a minimization problem:

$$v_i^{\text{opt}} = \text{argmin}_{v_i} \|V - V_h\|_F, \quad i = 1, ..., N$$

Hence, V_h is the projection of $V(p)$ over S_N. $\|.\|_F$ is a norm involving the state equation. In this approach, the quality of the solution is monitored either through the mesh size (i.e. $N \to \infty$) and/or the order of the finite element (i.e. $W_i \in P^m$ with m increasing for higher accuracy) [8]. If the approach is consistent, the projected solution tends to the exact solution when $N \to \infty$ or $m \to \infty$. In all cases, the size of the problem is large $1 \ll N < \infty$.

In a low-complexity approach, one approximate $V(p)$ by its projection $\tilde{V}$ over a subspace $\tilde{S}_n(\{w_i, i = 1, ..., n\})$ no longer generated by polynomial functions.

One rather considers $\{w_i, i = 1, ..., n\}$ a family of solutions (snapshots) of the initial full model $(p \to V(p))$:

$$\tilde{V} = \sum_{i=1}^{n} \tilde{v}_i^{\text{opt}} w_i,$$

$$v_i^{\text{opt}} = \text{argmin}_{v_i} \|V - \tilde{V}\|_F, \quad i = 1, ..., n.$$

The cost of the two methods depends on the cost of the solution of the minimization problems. For the reduced-order approach to be efficient, one aims therefore at $n \ll N$ [50]. This is only possible if the w_i family is well suited to the problem, in which case one can also expect a very accurate solution despite the small size of the problem. However, calculation of snapshots is not always easy and affordable especially when the original model involves a nonlinear partial differential equation. Reduced-order modeling can also be considered by removing the calculation of the snapshots w_i taking advantage of what we know on the physics of the problem and replacing the direct model $F(V(p)) = 0$ by an approximate model $f(V(p)) = 0$ for which snapshots are either analytically known or at least easier to evaluate. This can also be seen as changing the norm in the minimization problem above.

Considering a simpler state equation is a natural way to proceed, as often one does not need all the details about a given state. Also it is sufficient for the low-complexity model to have a local validity domain: one does not necessarily use the same low-complexity model over the whole range of the parameters. However, this brings in the question of the region of confidence for a given reduced-order model for which one needs to know the domain of validity.

7.6.1 Data interpolation

The most efficient approach to build a reduced-order model is to use data interpolation and regression to create a polynomial parametric model for the functional. One should keep in mind that in realistic configurations one has very little information about the functional. For instance, using a linear combination of observations (each observation is a call to the full model) gives:

$$V(x) \sim v(x) = \sum_{j=1}^{n_{\text{obs}}} \lambda_j(x) V_{\text{obs}}(x_j), \quad 0 \leq \lambda_j(x) \leq 1, \quad (7.16)$$

where $\lambda_j(x)$ are barycentric functions such that

$$\sum_{j=1}^{n_{\text{obs}}} \lambda_j(x) = 1, \quad \text{and} \quad \lambda_j(x_i) = \delta_{ij}.$$

To account for errors in observations V_{obs} a kriging construction [31, 12] can be used. If one assumes V to behave as a random variable, and supposing its

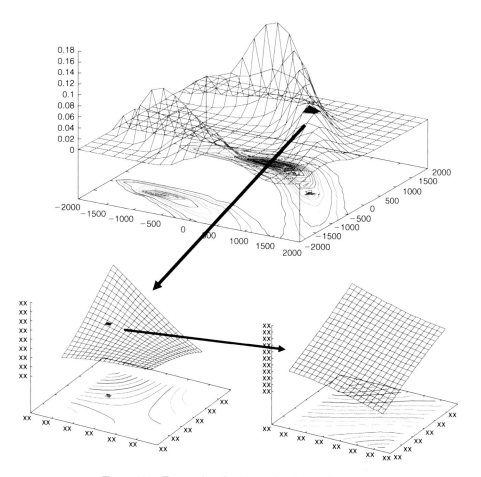

FIG. 7.6. Example of a three-level construction.

mean and covariance being those of V_{obs}, the kriging predictor v minimizes the variance of the prediction error: $\varepsilon = V - v$.

Hence, an interpolation using arbitrary regression polynomials ϕ of degree less than n_{deg} can be modified adding a kriging correlation term:

$$v(x) = \sum_{j=1}^{n_{\text{deg}}} a_j \phi_j(x) + \sum_{j=1}^{n_{\text{obs}}} b_j \psi(x, x_j), \quad (7.17)$$

where a_j are the regression coefficients, and b_j the kriging coefficients defined to minimize the variance of the prediction error. ψ is a Gaussian correlation function of the form:

$$\psi(x, x_j) = \exp\left(-\sum_{k=1}^{n_{\text{obs}}} \left(\frac{\|x^k - x_j^k\|^2}{r^k}\right)\right), \quad r^k > 0.$$

r^k defines the localization of the correction and variable r^k permits anisotropic interpolation.

7.6.1.1 Multi-level construction
As we said, the cost of direct simulation is prohibitive. One completes the reduced-order model only locally when possible. This is similar to mesh adaptation where nodes are added only where necessary.

The previous construction can be performed recursively on a cascade of embedded rectangular homothetic domains in the parameter space $\omega^i \subset \mathcal{O}_{ad}, i = 0, ...,$ with ω^0 the full search domain. [7] In this construction information is only transferred from coarse to fine levels from corners (see Fig. 7.6). Then calls for new observations are made where necessary to enrich the data basis. The refinement can use a quad tree [15] refinement algorithm or uses algorithms based on mesh adaptation as shown in Chapter 4. When the dimension of parameter space is large sparse grids should be used to avoid the curse of dimensionality [46, 6]. The variable v is then expressed as:

$$v = \sum_{i=0}^{n_{\text{level}}} v^i \chi(\omega^i) \tag{7.18}$$

where v^0 is the interpolation on the coarser level and $\chi(\omega^i)$ is the characteristic function for the subdomain on which level i is defined. The correction is zero outside ω^i. n_{level} is the total number of levels used. For $i > 1$, the restriction from level $i-1$ to i is evaluated using the information at the corners q_j of a brick (N being the size of the control space):

$$v^i(x) = \sum_{j=1}^{2^N} \lambda_j(x) v^{i-1}(q_j) \tag{7.19}$$

$$\sum_{j=1}^{2^N} \lambda_j(x) = 1, \quad \text{and} \quad \lambda_j(x_i) = \delta_{ij}, \quad i, j = 1, ..., 2^N$$

Figure 7.6 shows an example of such a construction for a functional with three levels of refinement and $N = 2$ [40, 5].

7.6.1.2 Uncertainties
We mentioned kriging to account for variability in observations. But this has limitations mainly related to knowledge of the variogram to establish the covariances.

---
[7] For the sake of simplicity, and also because this is often rich enough, we deliberately limit ourselves to rectangular configurations.

Let us describe another statistical approach to account for fluctuations in observations. For instance, consider the fluctuations in time of a functional $J(t)$ in the context of shape optimization for unsteady flows (also see Chapter 8). This is computed using a direct simulation loop linking the parameterization to the functional and involving a costly solution of the state equation.

Let us decompose J into a mean and a fluctuating part with zero mean:

$$J = \bar{J} + J', \quad \bar{J'} = 0,$$

where a time average is performed over the time interval of interest T:

$$\bar{J} = \frac{1}{T} \int_0^T J(t) dt.$$

If the flow is stationary $J' = 0$ and $J = \bar{J}$.

If perturbations are weak the deviation from the mean tendency is small and can be represented by a normal law for instance:

$$J' = \mathcal{N}(0, \sigma), \quad 0 \leq \sigma \ll 1.$$

An elegant way to account for small variations of observations is to take advantage of the low-complexity of the interpolation procedure (or any reduced-order modeling) and perform Monte Carlo estimation of the deviation.

Indeed, consider a set of multi-level observations (simulations) $j = 1, ..., n_{trials}$: $\{J^{i,j}, i = 1, ..., n_{level}\}$ where the trials are performed for admissible random choices of J^j through

$$J^j = \bar{J} + v^j, \quad v^j \in \{\mathcal{N}(0, \sigma)\}^2.$$

One can then define ensemble averages:

$$\bar{J^i} \sim \frac{1}{n_{trials}} \sum_{j=1}^{n_{trials}} J^{i,j}.$$

For a given level i one can have an estimation of the deviation from mean tendency for the velocity field

$$w^i = J^i - \bar{J^i},$$

and because $\bar{\bar{J^i}} = \bar{J^i}$, one has:

$$\bar{w^i} = 0,$$

with corresponding local standard deviations using, for instance, the maximum-likelihood estimate after assuming normal distribution for the results around their mean:

$$\sigma^i \sim \left(\int_{\omega^i} (w^i)^2 \right)^{1/2},$$

Beyond unsteadiness, this approach can be used to analyze the effect of any randomness or uncertainties in the definition of a control parameter (i.e. small fluctuations in flight incidence, etc.).

7.7 Optimal transport and shape optimization

Let us briefly introduce the problem of defining a shape from its curvature. This is fundamental in shape design in general because prescribing the curvature enters in the definition of developable surfaces [32]; it also leads to a class of optimization problems particularly difficult numerically and hence a good test of global optimization methods.

Developable surfaces are interesting in engineering design as they can be shaped from flat plates without stretching. The required forces are therefore significantly lower than for non-developable surfaces. Indeed, if one aims to build light objects with thin materials the internal forces necessary to conform the material to a non-developable surface can cause damage.

A developable surface is defined as a surface with zero Gauss curvature, which means that the product of the largest to smallest curvatures should vanish. On the other hand, at least one of the principal curvatures must be zero. Intuitively, a developable surface is a surface which can be rolled without slipping on a flat surface, making contact along a straight line, like a cylinder or a cone. In general, in design we aim at avoiding surfaces with negative Gauss curvature.

The mass transport problem was first considered by Monge in 1780 in his *Mémoire sur les remblais et les déblais* and revisited by Kantorovich in 1948 [28]. This problem has regained interest more recently as it models wide ranges of applications from image processing and physics to economics (see [51, 4, 7] for exhaustive references).

Numerical solution of the Monge-Ampère equation (MAE) with classical methods for PDEs is difficult as this equation changes type following the sign of its right-hand side (the PDE is elliptic if it is strictly positive, parabolic if it vanishes and hyperbolic if strictly negative).

Current numerics only target strictly positive right-hand-side and use either fixed point Newton type iterations [36] or minimization approaches for an augmented Lagrange formulation [13, 4]. We would like to solve the MAE via a minimization problem in the context of designing a surface from its curvature [39]. The optimal transport problem corresponds to the particular case of strictly positive curvature.

Consider two bounded, positive and integrable functions ρ_0 and ρ_1 with compact support in $\mathbf{R}^n$ such that

$$\int_{\mathbf{R}^n} \rho_0 = \int_{\mathbf{R}^n} \rho_1. \tag{7.20}$$

One would like to find the application $M : \mathbf{R}^n \to \mathbf{R}^n$ realizing the transformation from ρ_0 to ρ_1 minimizing, for instance,

$$\int_{\mathbf{R}^n} \|x - M(x)\|^2 \rho_0(x), \tag{7.21}$$

and such that for all continuous functions φ

$$\int_{\mathbf{R}^n} \rho_0(x)\varphi(M(x)) = \int_{\mathbf{R}^n} \rho_1(x)\varphi(x). \tag{7.22}$$

One knows that there is a unique $M = \nabla\psi$ with ψ a convex potential satisfying (7.21) [51]. In addition, if densities are strictly positive and continuous then one can look for ψ a solution of MAE ($H(\psi)$ being the Hessian of ψ and $\det(H)$ its determinant),

$$\det(H(\psi(x))) = \frac{\rho_0(x)}{\rho_1(\nabla\psi(x))}. \tag{7.23}$$

Solution to this problem can also come from a continuous time method minimizing some distance function between densities ρ_0 and ρ_1. The Wasserstein distance provides a solution for (7.23). It can be defined via a conservative flow:

$$d_W(\rho_0, \rho_1)^2 = \inf \int_{\mathbf{R}^n} \int_0^1 \rho(t,x) \|v(t,x)\|^2, \tag{7.24}$$

where (ρ, v) satisfies the following conservation law:

$$\rho_t + \nabla.(\rho v) = 0, \quad \rho(0,x) = \rho_0(x), \quad \rho(1,x) = \rho_1(x). \tag{7.25}$$

This is therefore a minimization problem with equality constraint. The optimality condition gives [4]

$$M = \nabla\psi = \nabla\phi + x, \tag{7.26}$$

where

$$\phi_t + 1/2|\nabla\phi|^2 = 0, \tag{7.27}$$

$$\rho_t + \nabla(\rho\nabla\phi) = 0, \quad \rho(0,x) = \rho_0(x), \quad \rho(1,x) = \rho_1(x), \tag{7.28}$$

with ϕ the Lagrange multiplier for the conservation constraint (7.25). This can be solved with an augmented Lagrange formulation minimizing the Eikonal equation residual under constraint (7.28) [4, 18, 13].

Here, we would like to go one step further and solve the case of a variable sign determinant in (7.23). Unlike in the positive sign case, the minimization problem for the residual of (7.23) cannot be solved using formulation (7.24-7.28).

In the sequel we consider the MAE for the definition of a 3D surface from a given curvature minimizing $J(\psi) = \|\psi_{xx}\psi_{yy} - \psi_{xy}^2 - f(x,y)\|$ where ψ is a given shape parameterization described in Chapter 6. To be sure of the existence of a solution, we build an inverse problem considering a target surface of the form $\psi(x,y) = (\sum_i a_i \sin(\omega_i x))(\sum_j b_j \sin(\omega_j y))$ with $a_i, b_j, \omega_{i,j} \in \mathbf{R}$. We build f from this surface and try to recover the surface. We would like to use our global minimization method to minimize the L^2 residual of MAE over a square calculation domain. The L^2 integral is discretized on a 10×10 Cartesian mesh corresponding to a control problem of dimension 100. Figure 7.7 shows a numerical solution of the MAE. The final and target surfaces match.

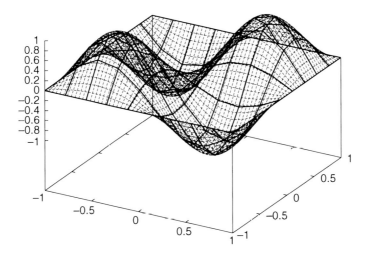

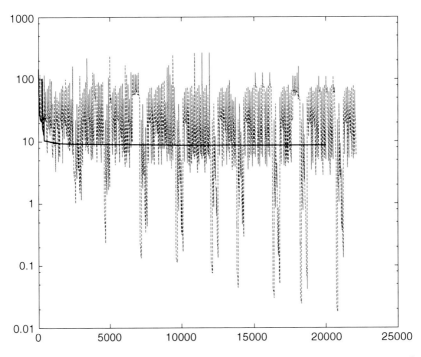

FIG. 7.7. Upper: example of surface recovery from given local curvature by a solution of the MAE for coarse and fine CAD-free parameterizations. Lower: convergence history showing visit of several local minima (right). One also shows the convergence to the quasi-Newton method used as the core minimization algorithm alone, captured by a local minimum.

References

[1] Attouch, H. Goudou, X. and Redont, P. (2000). The heavy ball with friction method, *Advances in Contemporary Math.* **2(1)**, 1-34.

[2] Attouch, H. and Cominetti, R. (1996). A dynamical approach to convex minimization coupling approximation with the steepest descent method, *J. Diff. Eq.*, **128(2)**, 519-540.

[3] Bock, H.G. and Plitt, K.J. (1984). A multiple shooting algorithm for direct solution of optimal control problems, *Proceedings of the 9th IFAC World Congress*, Budapest.

[4] Brenier, Y. and Benamou, J.-D. (2000). A computational fluid mechanics solution to the Monge-Kantarovich mass transfer problem, *Numerish Math.* **84(3)**, 375-393.

[5] Brun, J. and Mohammadi, B. (2007). Similitude généralisée et modélisation géométrique à complexité réduite, *CRAS*, **2511**, 634-638.

[6] Bungartz, H.-J. and Griebel, M. (2004). Sparse Grids, *Acta Numerica*, **13**, 147269.

[7] Carlier, G. and Santambrogio, F. (2005). A variational model for urban planning with traffic congestion, *ESAIM, COCV*, **11**, 595-613.

[8] Ciarlet, Ph. (1978). *The Finite Element Method for Elliptic Problems.* North-Holland, Amsterdam.

[9] Das, I. and Dennis J. E. (1997). A closer look at drawbacks of minimizing weighted sums of objectives for Pareto set generation in multicriteria optimization problems, *Struct. Optim.*, **14(1)**, 63-69.

[10] Das, I. and Dennis, J.E. (1998). Normal-boundary intersection: a new method for generating Pareto optimal points in multicriteria optimization problems, *SIAM J. Optim*, **8**, 631657.

[11] Deb, K. (1999). Multi-objective genetic algorithms: Problem difficulties and construction of test problems, *IEEE J. Evolut. Comp.*, **7**, 205-230.

[12] Chiles, JP. and Delfiner, P. (1989). *Geostatistics, Modeling Spatial Uncertainty,* John Wiley, London.

[13] Dean, E. and Glowinski, R. (2003). Numerical solution of the two-dimensional elliptic Monge-Ampère equation with Dirichlet boundary conditions: an augmented Lagrangian approach, *C. R. Acad. Sci. Paris*, **336**, 445-452.

[14] Dumas, L. Herbert, V. and Muyl, F. (2004). Hybrid method for aerodynamic shape optimization in automotive industry. *Comput. Fluids*, **33(5)**, 849858.

[15] Finkel, R. and Bentley, J.L. (1974). Quad trees: a data structure for retrieval on composite keys. *Acta Informatica*, **4(1)**, 1-9.

[16] Floudas, C.A. (2000). *Handbook of Test Problems in Local and Global Optimization*, Kluwer Academic, New York.

[17] Fonseca, C. and Fleming, J. (1995). An overview of evolutionary algorithms in multi-objective optimization. *Evolut. Comput.*, **3(1)**, 1-16.

[18] Fortin, M. and Glowinski, R. (1983). *Augmented Lagrangian Methods. Applications to the Numerical Solution of Boundary Value Problems,* Studies in Mathematics and its Applications, North-Holland, New York.

[19] Goldberg, D. (1989). *Genetic Algorithms in Search, Optimization and Machine Learning.* Addison Wesley, New York.
[20] Gorban, A. Kegl, B. Wunsch, D. and Zinovyev, A. (2007). *Principal Manifolds for Data Visualisation and Dimension Reduction*, LNCSE 58, Springer, Berlin.
[21] Hoel, P.G. (1971). *Introduction to Mathematical Statistics*, John Wiley, London.
[22] Ivorra, B. Hertzog, D. Mohammadi, B. and Santiago, J. (2006). Global optimization for the design of fast microfluidic protein folding devices, *Int. J. Num. Meth. Eng.* **26(6)**, 319333.
[23] Ivorra, B. Mohammadi, B. and Ramos, A. (2007). Semi-deterministic global optimization method and application to the control of Burgers equation, *JOTA*, **135(1)**, 54956.
[24] Ivorra, B. Mohammadi, B. and Ramos, A. (2008). Optimization strategies in credit portfolio management, *J. of Global Optim.*, **43(2)**, 415-427.
[25] Ivorra, B. Mohammadi, B. Redont, P. and Dumas, L. (2006). Semi-deterministic vs. genetic algorithms for global optimization of multichannel optical filters, *Int. J. Comput. Sci. Eng.*, **2(3-4)**, 170188.
[26] Jeong, S. Chiba, E. and Obayashi, S. (2005). Data mining for aerodynamic design space. *J. of Aerospace Comput., Inform. Comm.*, **2**, 218-223.
[27] Jolliffe, I.T. (2002). *Principal Component Analysis*, Springer Series in Statistics, 2nd edn., Springer, New York.
[28] Kantorovich, L.V. (1948). On a problem of Monge, *Uspelkchi Mat. Nauk.* **3**, 225-226.
[29] Kazuyuki, F. Fico, N. and de Mattos, B. (2006). Multi-point wing design of a transport aircraft using a genetic algorithm. *AIAA*, **2006-226**.
[30] Kohonen, T. (1995). *Self-Organizing Maps*, Springer, Berlin.
[31] Krige, D.G. (1951). *A statistical approach to some mine valuations and allied problems at the Witwatersrand.* Master thesis University of Witwatersran.
[32] Kuhnel, W. (2002). *Differential Geometry: Curves-Surfaces-Manifolds.* AMS, Providence.
[33] Kumano, T. Jeong, S. Obayashi, S. Ito, Y. Hatanaka, K. and Morino, H. (2006). Multidisciplinary design optimization of wing shape for a small jet aircraft using Kriging model. *AIAA*, **2006-932**.
[34] Kumano, T. Jeong, S. Obayashi, S. Ito, Y. Hatanaka, K. and Morino, H. (2005). Multidisciplinary design optimization of wing shape with nacelle and pylon. *Proc. Eccomas CFD-2006*, P. Wesseling, E. Onate, J. Périaux (Eds).
[35] Lindman, H.R. (1974). *Analysis of Variance in Complex Experimental Designs*, Freeman, New York.
[36] Loeper, G. and Rapetti, F. (2005). Numerical solution of the Monge-Ampere equation by a Newton's algorithm, *C. R. Acad. Sci. Paris*, **340**, 319-324.
[37] Mandic, D. and Chambers, J. (2001). *Recurrent Neural Networks for Prediction: Architectures, Learning Algorithms and Stability.* John Wiley, London.

[38] Messac, A. Sundararaj, G. J. Tappeta, R. and Renaud, J. E. (2000). Ability of objective functions to generate points on non-convex Pareto frontiers. *AIAA J.*, **38(6)**, 1084-1091.

[39] Mohammadi, B. (2007). Optimal transport, shape optimization and global minimization, *C. R. Acad. Sci. Paris*, **3511**, 591-596.

[40] Mohammadi, B. and Bozon, N. (2007). GIS-based atmospheric dispersion modeling, *Proc. FOSS4G Conference 2008, Cap Town*, ISBN 978-0-620-42117-1.

[41] Mohammadi, B. and Pironneau, O. (2004). Shape Optimization in Fluid Mechanics, *Ann. Rev. Fluid Mech.*, **36(1)**, 29-53.

[42] Mohammadi, B. and Redont, P. (2009). Global optimization and generation of Pareto fronts. *CRAS*, **347(5)**, 327-331.

[43] Mohammadi, B. and Saiac, J.-H. (2003). *Pratique de la Simulation Numérique*. Dunod, Paris.

[44] Pareto, V. (1906). *Manuale di Economia Politica*, Societa Editrice Libraria, Milano, Italy. Translated into English by Schwier, A.S. 1971: Manual of Political Economy, Macmillan, New York.

[45] Pediroda, V. Poloni, and C. Clarich, A. (2006). A fast and robust adaptive methodology for airfoil design under uncertainties based on game theory and self-organizing-pap theory. *AIAA*, **2006-1472**.

[46] Smolyak, SA. (1963). Quadrature and interpolation formulas for tensor products of certain classes of functions. *Dokl. Akad. Nauk SSSR*, **148**, 10421043. Russian, Engl. Transl.: Soviet Math. Dokl. **4**, 240243.

[47] Spooner, J. T. Maggiore, M. Onez, R. O. and Passino, K. M. (2002). *Stable Adaptive Control and Estimation for Nonlinear Systems: Neural and Fuzzy Approximator Techniques*. John Wiley, New York.

[48] Stadler, W. (1979). A survey of multicriteria optimization, or the vector maximum problem. *JOTA*, **29**, 1776-1960.

[49] Vanderplaats, G.N. (1990). *Numerical Optimization Techniques for Engineering Design*. Mc Graw-Hill, New York.

[50] Veroy, K. and Patera, A. (2005). Certified real-time solution of the parametrized steady incompressible Navier-Stokes equations: Rigorous reduced-basis a posteriori error bounds *Int. J. Numer. Meth. Fluids*, **47(8)**, 773-788.

[51] Vilani, C. (2003). *Topics in Optimal Transportation*, Graduate Studies in Mathematics series, **58**, AMS.

[52] Zingg, D. and Elias, S. (2006). On aerodynamic optimization under a range of operating conditions. *AIAA*, **2006-1053**.

8
INCOMPLETE SENSITIVITIES

8.1 Introduction

This chapter is devoted to the concept of incomplete sensitivity in shape optimization. The aim is to avoid when possible the linearization of the state equation. We will see that sometimes this goes through a reformulation of the initial problem.

For the design of a shape S, consider a general situation with a geometrical design or control variable x fixing $S(x)$, auxiliary geometrical parameters $q(x)$ (mesh related informations), a state or flow variable $U(q(x))$ and a criterion or cost function for optimization J

$$J(x) : x \to q(x) \to U(q(x)) \to J(x, q(x), U(q(x))). \tag{8.1}$$

The derivative of J with respect to x is

$$\frac{dJ}{dx} = \frac{\partial J}{\partial x} + \frac{\partial J}{\partial q}\frac{\partial q}{\partial x} + \frac{\partial J}{\partial U}\frac{\partial U}{\partial x}. \tag{8.2}$$

The major part of the computing time of this evaluation is due to $\partial U/\partial x$ in the last term (see also Chapter 5).

We have observed that when the following requirements are satisfied, the last term in (8.2) is small

1. J is of the form $J(x) = \int_S f(x, q(x))g(U)$;
2. the local curvature of S is not too large (this needs to be quantified for each case, for a wing typically this concerns regions away from leading and trailing edges.)

If these requirements are met, our experience shows that an incomplete evaluation of the gradient of J can be used, neglecting the sensitivity with respect to the state U in (8.2). This does not mean that a precise evaluation of the state is not necessary, but that for a small change in the shape the state will remain almost unchanged, while geometrical quantities, such as the normal to S, for instance, have variations much greater in magnitude. An optimization method using incomplete sensitivity is a sub-optimal control method and in that sense has limitations.

Historically, we discovered the incomplete sensitivity concept after manipulating automatic differentiation tools.[8] We had the difficulty that the generated

---

[8] The Odyssee tool at this time [4,43,10] but what will be said here is valid for any AD tool.

codes were too big to be efficiently handled; and this especially in reverse mode necessary in shape optimization. We then tried to see if it was possible to drop some parts of the direct loop and observed that for the class of functionals mentioned above, the sensitivity was mainly geometric.

8.2 Efficiency with AD

In this section, we give some advice for an efficient use of AD. The objective is to reduce the computing cost of sensitivities for some important functions.

8.2.1 Limitations when using AD

Suppose one has an automatic differentiator capable of producing gradients or Jacobians and a differentiator in reverse mode capable of computing cotangent linear applications (see Chapter 5). The limitation of the reverse mode comes from the memory required to store Lagrange multipliers and intermediate variables. The need for intermediate storage can be explained by the following:

If n is the number of controls, let $f : R^n \to R$ such that

$$f(x) = f_3 o f_2 o f_1 o f_0(x).$$

where f_0 smoothes the controls x, f_1 defines the mesh points in terms of x, f_2 computes the flow and f_3 evaluates the cost. The reverse mode produces the Jacobian $D^t f : R \to R^n$, as

$$D_x^t f = D_x^t f_0 D_{f_0(x)}^t f_1 D_{f_1 o f_0(x)}^t f_2 D_{f_2 o f_1 o f_0(x)}^t f_3.$$

We can see that we have to store (or recompute) the intermediate states (i.e. $f_0(x), f_1 o f_0(x)$ and $f_2 o f_1 o f_0(x)$) before making the product.

This can be a real problem when using an iterative method (for time integration for instance). Indeed, in this case, the intermediate states cannot be recomputed as the cost will grow as the square of the number of iterations. Therefore, they have to be stored somehow.

To give a more precise idea, consider our flow solver which is explicit with one external loop for time integration of size KT and internal nested loops.

```
External Loop in time 1,...,KT

  Loop over triangles (tetrahedra) 1,..,NT
    NAT affectations
  End loop NT

  Loop over edges    1,..,NE
    NAE affectations
  End loop NE

  Loop over nodes    1,..,NN
```

```
    NAN affectations
  End loop NN

End external loop KT
  cost evaluation
```

Inside each time step, we have loops on nodes, segments, and tetrahedra (triangles in 2D) of sizes NN, NS, and NT. Inside each internal loop, we have affectations like:

`new_var = expression(old_var)`

describing the spatial discretization. The number of these affectations are NAN, NAS and NAT. The required memory to store all the intermediate variables is therefore given by:

$$M = KT \times ((NN \times NAN) + (NS \times NAS) + (NT \times NAT)).$$

This is out of reach even for quite coarse meshes. To give an idea, for the 3D configuration presented here, we have:

$$KT \sim 10^4, \quad (NN, NS, NT) \sim 10^5, \quad (NAN, NAS, NAT) \sim 10^2,$$

which makes $M \sim 10^{11}$ words. We can use an implicit method to reduce the number of time steps and therefore the storage. One alternative here is to use the check-pointing scheme of Griewank [5] to have a base 2 logarithmic growth of the complexity of the algorithm or use a compressive sampling technique [1] to minimize the number of samples needed for any state recovery or more generally meta-models or reduced order models to again break the complexity.

8.2.2 Storage strategies

To understand the need for a storage strategy, let us look at what we get if we store all the intermediate states, no intermediate state, and some of them.

Full storage case KT states stored, KT forward iterations (i.e. one forward integration) and KT backward iterations (i.e. one backward integration).

No storage case 0 states stored, $(KT + KT * (KT - 1)/2)$ forward iterations (i.e. one forward integration plus computing the required state during the backward integration each time) and KT backward iterations (i.e. one backward integration).

Intermediate storage n states stored non-uniformly over KT states (with $n \ll KT$), $(KT + \Sigma_{i=1}^{n}(m_i*(m_i-1)/2))$ forward iterations (i.e. one forward integration plus computing the required state each time starting from the nearest stored state, m_i being the number of iterations between two closest stored states) and KT backward iterations (i.e. one backward integration). Of course, if the stored states are uniformly distributed, we have $(KT + n(m * (m - 1)/2))$ forward iterations, so with one state stored ($n = 1$), we reduce by a factor of 2 the extra work ($m = KT/2$).

We can see that in each case the backward integration has been done only once.

8.2.3 Key points when using AD

Here are some remarks coming from the physics of the problem.

8.2.3.1 Steady flows
The first key remark concerns steady target applications, in the sense that the history of the computation is not of interest. This history is often evaluated using a low accuracy method and local time stepping. It is obvious therefore that it is not necessary to store these intermediate states. In other words, when computing the gradient by reverse mode, it is sufficient to store only one state, if we start with an initial state corresponding to the steady state for a given shape. This reduces the storage by the number of time steps (KT).

$$M = ((NN \times NAN) + (NS \times NAS) + (NT \times NAT)).$$

8.2.3.2 Inter-procedural differentiation
The second important point is to use inter-procedural [10] derivation. The idea is to use dynamic memory allocation in the code. We replace what is inside an internal loop by a subroutine and differentiate this subroutine. This will reduce the required memory to:

$$M = NAN + NAS + NAT,$$

but will imply extra calls to subroutines [10]. In fact, this is what we need if the memory is declared as static, and we need less if we choose to re-allocate the memory needed by subroutines.

8.2.3.3 Accuracy of the adjoint state for steady targets
Here again we assume that a steady state can be reached by the time-dependent solver of the state equation. By construction the adjoint state equation is also time-dependent. We saw that, when starting from the steady state for a given shape, we only need one iteration of the forward procedure before starting the reverse integration. Therefore, we need to know what accuracy for convergence is sufficient for the adjoint for the gradient algorithm to converge. In other words, how can we choose the maximum time for the time integrator of the adjoint?

We denote by J^∞ the value of the cost function of the optimization problem at convergence and by J^n the value of the cost function at step n. We notice that at convergence we would like to have $\nabla J_h^\infty = 0$. This will be achieved if for some $\beta \in (0,1)$ $|\nabla J_h^n| < \beta |\nabla J_h^{n-1}|$. This gives a criterion to stop the reverse time step loop. In other words, we have two different numbers of time step, one for the forward system (say one) and another for the adjoint system (very large), and we leave the reverse time integration loop once the above criterion is satisfied. In the previous code, we therefore need to evaluate the cost function inside the time integration loop, but this is cheap.

8.2.3.4 Localized gradient computation
This remark can be understood from Fig. 8.1. We can see that the variations of the internal mesh nodes have no effect

on the cost function. Therefore, when computing the gradient, the internal loops presented above on nodes, segments, and elements are only computed on a few element layers around the shape.

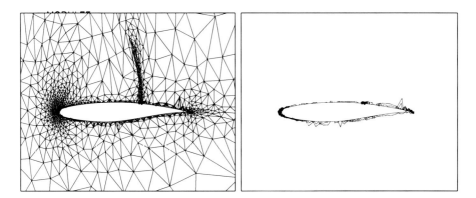

FIG. 8.1. $\partial J/\partial x_{mesh}$. The gradient is concentrated along the first element layer along the geometry. This is a key remark to reduce the computational effort.

8.3 Incomplete sensitivity

We noticed that the previous remark was mainly valid for cost functions based on information about the shape (for instance through boundary integrals). Fortunately, this is often the case for applications. Therefore, when the cost function is based on local information around the shape, the sensitivity of the cost function with respect to the state can be neglected. This does not mean that a precise evaluation of the state is not necessary, but that for a small change in the shape the state will remain almost unchanged, while geometrical quantities have variations of the same order:

$$\delta x \sim |\delta n| \sim O(1), \quad \text{but} \quad \delta u \sim O(\varepsilon).$$

8.3.1 Equivalent boundary condition

We recover here the argument behind effective transpiration conditions widely used to represent small shape deformations. More precisely, consider the sensitivity of the product $\vec{u}.\vec{n}$ with respect to the shape parameterization x. Formally, we have:

$$\frac{d}{dx}(\vec{u}.\vec{n}) = \frac{d\vec{u}}{dx}.\vec{n} + \vec{u}\frac{d\vec{n}}{dx} \sim \vec{u}\frac{d\vec{n}}{dx}.$$

Indeed, the argument above means that $|d\vec{u}/dx| \ll |d\vec{n}/dx|$. We notice that u has to be accurate for an accurate sensitivity.

8.3.2 Examples with linear state equations

Recall that the idea behind incomplete gradients is that the changes in the state are negligible compared to those of the geometry for small variations of the domain.

Consider as cost function $J = \epsilon^n u_x(\epsilon)$ and as state equation the following Poisson equation (taking $|\epsilon| \ll 1$)

$$-u_{xx} = 1, \quad \text{on }]\epsilon, 1[, \quad u(\epsilon) = 0, \quad u(1) = 0,$$

which has as solution $u(x) = -x^2/2 + (\epsilon+1)x/2 - \epsilon/2$. This is a case of a function which has a strong geometrical element and a weak dependance on the geometry via the state u. The gradient of J with respect to ϵ is given by

$$J_{,\epsilon}(\epsilon) = \epsilon^{n-1}(nu_x(\epsilon) + \epsilon u_{x\epsilon}(\epsilon)) = \frac{\epsilon^{n-1}}{2}(-n(\epsilon-1) - \epsilon).$$

The second term between parentheses, $-\epsilon$, is the state linearization contribution which is neglected in incomplete sensitivities. We can see that the sign of the gradient is always correct and the approximation is better for large n.

Now, let us make the same analysis for an advection diffusion equation with Pecley number Pe with the same cost function:

$$u_x - Pe^{-1} u_{xx} = 0, \quad \text{on }]\epsilon, 1[, \quad u(\epsilon) = 0, \quad u(1) = 1.$$

The solution of this equation is

$$u(x) = \frac{\exp(Pe\ \epsilon) - \exp(Pe\ x)}{\exp(Pe\ \epsilon) - \exp(Pe)}. \tag{8.3}$$

Hence

$$u_x(x) = \frac{-Pe\ \exp(Pe\ x)}{\exp(Pe\ \epsilon) - \exp(Pe)},$$
$$(u_x)_{,\epsilon}(x = \epsilon) = \frac{(Pe\ \exp(Pe\ \epsilon))^2}{(\exp(Pe\ \epsilon) - \exp(Pe))^2} = u_x^2(x = \epsilon).$$
$$J_{,\epsilon}(x = \epsilon) = \epsilon^{n-1} u_x(\epsilon)(n + \epsilon u_x(\epsilon)).$$

Therefore, if $|\epsilon u_x(\epsilon)| \ll n$ the contribution coming from state linearization can be neglected.

The previous analysis holds for any functional of the form $J = f(\epsilon)g(u, u_x)$.

Another example concerns Poiseuille flow in a channel driven by a constant pressure gradient (p_x). The walls are at $y = \pm S$. The flow velocity satisfies:

$$u_{yy} = \frac{p_x}{\nu}, \quad u(-S) = u(S) = 0. \tag{8.4}$$

The analytical solution satisfying the boundary conditions is $u(S, y) = \frac{p_x}{2\nu}(y^2 - S^2)$.

Consider the sensitivity of the flow rate with respect to the channel thickness. The flow rate is given by $J_1(S) = \int_{-S}^{S} u(S,y)dy$. The gradient is given by (using the boundary conditions in (8.4)):

$$\frac{dJ_1}{dS} = \int_{-S}^{S} \partial_S U(S,y)dy = \frac{-2S^2 p_x}{\nu},$$

while the incomplete sensitivity (dropping the state derivative) vanishes.

Now consider

$$J_2(S) = S^n J_1(S).$$

The gradient of the cost function with respect to the control variable S is:

$$\frac{dJ_2}{dS} = nS^{n-1} J_1(S) + S^n \frac{dJ_1}{dS}.$$

We can see now that we have two terms involved: the first integral is what the incomplete sensitivity gives and the second comes from the linearization of the state:

$$\frac{dJ_2}{dS} = -\frac{4nS^{n+2} p_x}{6\nu} - \frac{S^{n+2} p_x}{\nu}.$$

We can see that the two contributions have the same sign and are of the same order, of course the geometrical sensitivity dominates for large values of n. This remark is interesting for axisymmetric applications in nozzles, for instance.

Now consider the application of incomplete sensitivities with a Burgers equation for the state already saw in Chapter 5:

$$u_t + 0.5(u^2)_x = \mu x u, \quad \text{on }]-1+S, 1[, \tag{8.5}$$

with

$$u(-1+S) = 1, \quad u(1) = -0.8, \quad |S| \ll 1.$$

We consider the steady solution of (8.5) and take the variations S on the left-hand side frontier as control parameter. Again, suppose the functional involves control times state quantities, for instance as in: $J(S) = Su_x(-1+S)$, the gradient is given by

$$J_S = u_x(-1+S) + Su_{xS}(-1+S). \tag{8.6}$$

Without computing the solution, it is clear from the equation that in regions where the solution is regular $u_x = \mu x$. The exact gradient is therefore $J_S = \mu(-1+S) + S\mu$ to be compared with the incomplete gradient $\mu(-1+S)$. We see that the sign of the incomplete gradient is always correct and the difference is negligible ($|S| \ll 1$).

8.3.3 Geometric pressure estimation

Designing a shape with minimum drag with respect to a parameter x (scalar for clarity) involves an integral on the shape of $p(x)u_\infty \cdot n(x)$:

$$C_d = \int_{\text{shape}} p(x) u_\infty \cdot n(x) ds.$$

Suppose the pressure is given by the Newton formula (also called the cosine-square law) $p = p_\infty (u_\infty \cdot n)^2$. We therefore have $pu_\infty n = p_\infty (u_\infty \cdot n)^3$. Its derivative with respect to x is

$$\frac{d(pu_\infty \cdot n)}{dx} = (pu_\infty) \cdot \frac{dn}{dx} + \frac{dp}{dx}(u_\infty \cdot n) = 3 p_\infty u_\infty (u_\infty \cdot n)^2 \frac{dn}{dx}.$$

On the other hand, incomplete sensitivity gives

$$\frac{d(pu_\infty n)}{dx} \approx (pu_\infty) \frac{dn}{dx} = p_\infty u_\infty (u_\infty n)^2 \frac{dn}{dx}.$$

We see that the two derivatives only differ by a factor of 3 and have the same sign. The expression above can be rewritten as $pu_\infty.n = p|u_\infty| \cos(\frac{u_\infty}{|u_\infty|}.n)$. The incomplete gradient is therefore $p(u_\infty.n)_x = -p|u_\infty| \sin(\frac{u_\infty}{|u_\infty|}.n) = 0$ when n is aligned with u_∞. The incomplete sensitivity fails therefore for these area (e.g. the area near the leading edge for instance for an airfoil at no incidence). In the same way, if we were interested in the evaluation of the lift sensitivity, the incomplete sensitivity would be wrong where n is close to $u_\infty^\perp$ (e.g. along the upper and lower surfaces of a wing at no incidence). This means that the incomplete sensitivity is not suitable for lift sensitivity evaluation, except if the deformation is along the normal to the shape as in that case the $(\partial p/\partial n = 0)$ boundary condition would imply that the incomplete and exact gradients are the same. This is also why we advocate the use of deformations normal to the wall as parameterization.

This is a worse case analysis, however, as linearization leads to equal order sensitivities for small variations of the local normal to the shape and the inflow velocity. Indeed, we know that small changes in the geometry in a high curvature area (e.g. leading and trailing edges) have important effects on the flow, much more than changes in area where the shape is flat.

A final remark on the accuracy of the state. Consider formally the linearization of the drag $C_d = pu_\infty.n$ with respect to x and its approximation:

$$\frac{d}{dx}(pu_\infty.n) = \frac{dp}{dx}(u_\infty.n) + p\frac{dn}{dx}.u_\infty \sim p\frac{dn}{dx}.u_\infty. \tag{8.7}$$

We notice that p (the state) has to be accurate even if used in an approximate calculation of the sensitivity. This shows, for instance, the importance of including viscous effects in the state evaluation for optimization [8] but not necessarily to take them into account in the sensitivities (see also Chapter 2). It also gives some insight into meshing issues as discussed below.

8.3.4 Wall functions

We saw above that with a homogeneous normal gradient boundary condition on the pressure $\partial p/\partial n = 0$ on the shape the incomplete and exact sensitivities coincide for shape variations normal to the wall. For viscous flow, the boundary condition for the velocity is no-slip at the wall. In addition, we saw in Chapter 3 wall functions such as:

$$u = u_\tau f(y^+), \quad y^+ = \frac{(y - y_w)u_\tau}{\nu}$$

describing the behavior of the tangential component u of the velocity near the shape with respect to the distance $y - y_w$ to it and with $f(0) = 0$. f is for instance the Reichardt law (3.60). Hence, one can express the solution in the domain as (denote the shape by y_w):

$$u = w(y - y_w)v(w(y - y_w))$$

where w tends to zero with the distance to the shape $y - y_w$ and v is selected to satisfy the flow equations. The wall function analysis suggests that such a decomposition exists. Now, sensitivity analysis for a functional such as $J = J(y_w, u)$ gives:

$$\frac{dJ}{dy_w} = \frac{\partial J}{\partial y_w} + \frac{\partial J}{\partial u}\frac{\partial u}{\partial y_w} = \frac{\partial J}{\partial y_w} + \frac{\partial J}{\partial u}(w\frac{\partial v}{\partial y_w} + v\frac{\partial w}{\partial y_w})$$

But, $w(0) = 0$ and one knows the dependency between w and y_w. Therefore, in cases where the near-wall dependency of the solution with respect to the distance to the shape is known the sensitivity with respect to shape variations normal to the wall can be obtained without linearizing the state equation. A similar analysis can be made for the temperature when an isothermal boundary condition is used (see Chapter 3) while with adiabatic walls the remark made for the pressure applies.

8.3.5 Multi-level construction

From expression (8.7) above it is clear that it is better to have a state computed with good precision and an approximate gradient than to try to compute an accurate gradient based on an incorrect state such as those computed on a coarse mesh. For instance if p_h, p_H denote a coarse and accurate computation, we claim that when incomplete gradients work

$$\cdot|\frac{d}{dx}(p\cdot n) - p_h\frac{dn}{dx}| \approx |\frac{d}{dx}(p\cdot n) - \frac{d(p_h\cdot n)}{dx}| \leq |\frac{d}{dx}(p\cdot n) - \frac{d(p_H\cdot n)}{dx}|.$$

The left-hand side is the difference between the exact and incomplete gradient computed on a fine mesh.

This error is often present and is due to the fact that the cost of iterative minimization and gradient evaluations force the use of meshes coarser than what would have been used for a pure simulation.

One possibility to avoid this difficulty is to use different meshes for the state and the gradient. This is the idea behind multi-level shape optimization where the gradient is computed on the coarse level of a multi-grid construction and where the state comes from the finer level:

$$\frac{d}{dx}(p \cdot n)_h \approx I_H^h \left(\frac{dp}{dx}\right)_H \cdot n + p_h \cdot \frac{dn}{dx}.$$

The first term on the right-hand side is the interpolation of the gradient computed on the coarse grid over the fine level.

These remarks are essential, as usually users are penalized by the cost of sensitivity evaluations which forces them to use coarser meshes in optimization than the meshes they can afford for the direct simulation alone.

8.3.6 Reduced order models and incomplete sensitivities

Note that in a computer implementation we can always try incomplete sensitivity, check that the cost function decreases and if it does not add the missing term. A middle path to improve on incomplete evaluation of sensitivities is to use a reduced complexity (or reduced order) model which provides an inexpensive approximation of the missing term in (8.2).

For instance, consider the following reduced model for the definition of $\tilde{U}(x) \sim U(x)$ (e.g. $\tilde{U}$ is the Newton formula for the pressure and U the pressure from the Euler system). Consider the following approximate simulation loop:

$$x \to q(x) \to \tilde{U}(q(x)). \tag{8.8}$$

The incomplete gradient of J with respect to x can be improved by evaluating the former term in (8.2) linearizing (8.8) instead of (8.1)

$$\frac{dJ}{dx} \sim \frac{d\tilde{J}}{dx} = \frac{\partial J(U)}{\partial x} + \frac{\partial J(U)}{\partial q}\frac{\partial q}{\partial x} + \frac{\partial J(U)}{\partial U}\frac{\partial \tilde{U}(U)}{\partial x}, \tag{8.9}$$

$\tilde{U}$ being an approximation of U is used here only to simplify the computation of $\partial \tilde{U}/\partial x$. It is important to notice that the reduced model needs to be valid only over the support of the control parameters. This approximation can be improved by considering the following chain:

$$x \to q(x) \to \tilde{U}(q(x))\left(\frac{U(x)}{\tilde{U}(x)}\right). \tag{8.10}$$

Linearizing (8.10) instead of (8.1) and freezing $U/\tilde{U}$ gives

$$\frac{dJ}{dx} \sim \frac{d\tilde{J}}{dx} = \frac{\partial J(U)}{\partial x} + \frac{\partial J(U)}{\partial q}\frac{\partial q}{\partial x} + \frac{\partial J(U)}{\partial U}\frac{\partial \tilde{U}}{\partial x}\frac{U(x)}{\tilde{U}(x)}. \tag{8.11}$$

A simple example shows the importance of the scaling introduced in (8.10). Consider $U = \log(1+x)$ and $J = U^2$. Naturally

$$dJ/dx = 2UU' = 2\log(1+x)/(1+x) \sim 2\log(1+x)(1-x+x^2+...).$$

Consider $\tilde{U} = x$ as the reduced complexity model, valid around $x = 0$. Then (8.9) gives $J' \sim 2U\tilde{U}' = 2\log(1+x)$ while (8.11) gives

$$J' \sim 2U\tilde{U}'(U/\tilde{U}) = 2\log(1+x)(\log(1+x)/x) \sim 2\log(1+x)(1-x/2+x^2/3+...)$$

which is a better approximation for the gradient.

The reduced order modeling above is also useful if we want to use descent methods with a coding complexity comparable to minimization algorithms such as genetic or simplex (i.e. only requiring the functional from the user), but with a much lower calculation complexity. With the reduced order modelling above finite differences or complex variable methods introduced in Chapter 5 again become viable for sensitivity analysis in large dimension optimization. For instance, we consider the following expression for finite differences:

$$\frac{dJ}{dx}(i) = \frac{J(x+\varepsilon e_i, U(x+\varepsilon e_i)) - J(x,U(x))}{\varepsilon}$$

$$\sim \frac{J(x+\varepsilon e_i, \tilde{U}(x+\varepsilon e_i)(\frac{U(x)}{\tilde{U}(x)})) - J(x,U(x))}{\varepsilon},$$

where $x+\varepsilon e_i$ indicates a small change in the ith component of x.

8.3.7 Redefinition of cost functions

To be able to use incomplete sensitivities we need to be in their domain of validity: mainly surface functionals on the unknown shape and involving products of state functions with functions of the geometry. This is the case for cost functions based on aerodynamic coefficients. But, if we consider an inverse design problem,

$$J(x) = \frac{1}{2}\int_\Gamma (f(u) - f_{\text{target}})^2 ds,$$

the cost function does not depend explicitly on the geometry. For these problems, we use the following modified cost function to calculate the incomplete sensitivity:

$$J(x) = \frac{1}{2}\int_\Gamma (f(u) - f_{\text{target}})^2 \left(\left(\frac{\vec{U}_\infty + \vec{U}_\infty^\perp}{||\vec{U}_\infty||}\right).\vec{n}\right)^n ds, \quad n \geq 2.$$

The decomposition along $\vec{U}_\infty$ and $\vec{U}_\infty^\perp$ is there to avoid the fact that the incomplete sensitivity vanishes in regions where the normal to the shape is normal to the inflow $(\cos(\theta)' \sim 0$ if $|\theta| \ll 1)$. Hence, one can recover some information on the sensitivity even for these regions.

We show several examples of cost functions and constraint redefinition in Chapter 10 for blade optimization, sonic boom and wave drag reductions and heat transfer optimization. Of course, as for the reduced complexity models in section 8.3.6, this redefinition is only considered for gradient evaluation.

8.3.8 Multi-criteria problems

For multiple criteria, where computing time is even more critical, incomplete sensitivity has an edge over full gradient calculation. When several functionals j_i are involved in a design problem (for instance drag reduction under lift and volume constraints as shown below):

$$\min_x j_1(u(x)), \quad \text{such that} \quad j_j(u(x)) = 0, \quad j = 2, ..., n,$$

we prefer to take advantage of the low-complexity of incomplete sensitivity on the Pareto functional:

$$J = \sum_i \alpha_i j_i.$$

To get J', we evaluate individual incomplete sensitivities j'_i and use a projection over the subspace orthogonal to constraints: $\{(..., j'^\perp_j, ...), j \neq i\}$. For instance, one can use:

$$\tilde{j}'_i = j'_i - \sum_{j \neq i} (j'_i, j'_j) j'_j;$$

then $J' = \tilde{j}'_{i_{max}}$ where $i_{max} = Argmin_i \|\tilde{j}'_i\|$.

This also makes possible the use of interior point methods requiring individual sensitivities. With a full gradient, this would have implied calculating an adjoint variable for each of the constraint.

8.3.9 Incomplete sensitivities and the Hessian

An a posteriori validation of incomplete sensitivities can be obtained with quasi-Newton algorithms working with the Hessians of the functionals [3]. A steepest descent method can be successful if the incomplete gradient is in the right half-space, while with BFGS, for instance, we need not only a correct direction but also some consistency in the amplitude of the gradient. Otherwise, the eigenvalues of the Hessian matrix might have the wrong sign.

Here we show one example of shape optimization comparing BFGS and steepest descent (SD) algorithms (see Fig. 8.3). We want to minimize drag and at the same time increase the lift coefficient and the volume:

$$\min C_d, \quad C_l^0 \leq C_l, \quad V^0 \leq V, \tag{8.12}$$

where C_d and C_l are, respectively, the aerodynamic drag and lift coefficients and V the volume (0 indicates the corresponding value for the initial configuration). Incomplete sensitivities can be used for these quantities as they are all boundary integrals. Indeed, the volume can be transformed into a boundary integral by denoting $\vec{X} = (x, y, z)^t$:

$$V = \int_\Omega 1 dv = \int_\Omega \frac{1}{3} \nabla.(\vec{X}) dv = \int_{\partial\Omega} \vec{X}.\vec{n} ds.$$

The initial geometry is an AGARD wing with an aspect ratio of 3.3 and a sweep angle of 45 degrees at the leading edge line. The computational mesh has 120

000 tetrahedra, 2800 grid points being located on the surface of the wing. The dimension of the design space is therefore 2800. The upper surface grid for the baseline configuration is shown in Fig. 8.2. The flow is characterized by a free-stream Mach number of 2 and an angle of incidence of 2°. The initial values of the lift coefficient, C_l^0, of the drag coefficient, C_d^0, and of the volume, V^0, are 0.04850, 0.00697, and 0.030545, respectively. Figure 8.2 shows the convergence history of the objective function versus the time required by the BFGS and the SD methods. Hessian computation requires 0.027 time units defined by the time the flow solver requires to drop the residual by three orders of magnitude. This takes nearly 300 Runge-Kutta iterations with a CFL of 1.3. The final values for the coefficients above are with the BFGS method ($C_l = 0.0495, Cd = 0.005, V = 0.03048$) and with the SD method ($Cl = 0.1473, Cd = 0.0188, V = 0.06018$). Hence, the drag coefficient is reduced by 28%, the lift coefficient increased by 2% and the volume almost unchanged.

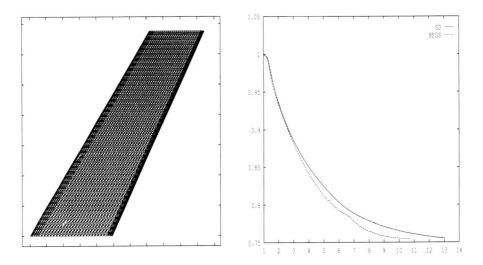

FIG. 8.2. AGARD wing. Upper surface discretization (left). Convergence for the objective function (right). One unit of work corresponds to the time to converge one single flow analysis.

8.4 Time-dependent flows

An important issue in using incomplete sensitivities is for time-dependent applications to enable for real-time sensitivity definition in the sense that the state and the sensitivities are available simultaneously without the need for solving an adjoint problem. This avoids the difficulty of intermediate state storage for time-dependent problems. Indeed, unlike in steady applications where intermediate states can be replaced by the converged state as mentioned in section 8.2.3.1,

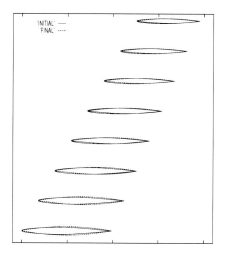 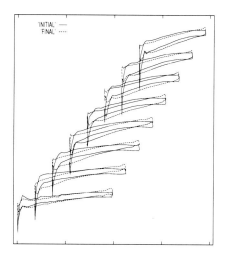

FIG. 8.3. AGARD wing. Initial and optimized wing shapes (left), BFGS and SD gives the same answer. Initial and optimized pressure coefficient distributions (right).

reducing the storage to one state, in time-dependent problems one cannot make this simplification.

We distinguish two situations: when the control is time-independent and when the control is time-dependent. A shape optimization problem for unsteady flows is in the first class, while a flow control problem with an injection/suction device belongs to the second domain.

Let us start with shape optimization for unsteady flows where the control (shape parameterization) does not depend on time even though the state and the cost functional are time-dependent.

The problem of shape optimization for unsteady flows can be formulated as:

$$\min_{S \in O_{ad}} J(S, \{u(t, q(S)), t \in [0, T]\}) \qquad (8.13)$$

where the state $u(t, q(S))$ varies in time but not S. The cost function involves state evolution over a given time interval $[0, T]$ through for instance:

$$J(S) = \frac{1}{T} \int_0^T j(S, q(S), u(t, q(S))) dt \qquad (8.14)$$

where j involves, for instance, instantaneous pressure based lift or drag coefficients:

$$C_d(t) = \frac{1}{\rho_\infty |\vec{u}_\infty|^2} \int_S (\vec{u}_\infty \cdot \vec{n}) p(t, q(S)) ds, \quad C_l(t) = \frac{1}{\rho_\infty |\vec{u}_\infty|^2} \int_S (\vec{u}_\infty^\perp \cdot \vec{n}) p(t, q(S)) ds$$

where ρ_∞ and $\vec{u}_\infty$ denote the reference density and velocity vector, taken for external flows as far field quantities, for instance.

The gradient of J is the averaged instantaneous gradients:

$$J'(S) = \frac{1}{T}\int_0^T j'(S,q(S),u(t,q(S)))dt = \frac{1}{T}\int_0^T (j_S + j_q q_S) + \frac{1}{T}\int_0^T j_u u_S.$$

Therefore, we only need to accumulate the gradient over the period $[0, T]$. The first term is the incomplete sensitivity. In case the full gradient is required then an adjoint problem is required to compute the remaining terms.

Shape optimization for unsteady flows has numerous applications. For instance, noise reduction as the radiated noise is linked to lift and drag time fluctuations [6]. For these problems we may minimize:

$$J(S) = \frac{1}{T}\int_0^T \left[\left(\frac{\partial C_d}{\partial t}\right)^2 + \left(\frac{\partial C_l}{\partial t}\right)^2 \right] dt,$$

where

$$(C_d)_t = \frac{1}{\rho_\infty |\vec{u}_\infty|^2} \int_S (\vec{u}_\infty . \vec{n}) \frac{\partial p(S,t)}{\partial t} ds, \quad (C_l)_t = \frac{1}{\rho_\infty |\vec{u}_\infty|^2} \int_S (\vec{u}_\infty^\perp . \vec{n}) \frac{\partial p(S,t)}{\partial t} ds. \tag{8.15}$$

Figure 8.4 shows a shape optimization problem for aerodynamic fluctuation reduction.

8.4.1 Model problem

To illustrate the above statements let us compute the sensitivities on a simple time-dependent model problem.

Consider the following time-dependent state equation for $u(y,t)$, $-S \leq y \leq S$, $t \geq 0$ in an infinite channel of width $2S$:

$$u_t - u_{yy} = F(S,y,t), \quad u(S,t) = u(-S,t) = 0, \tag{8.16}$$

with

$$F(S,y,t) = -\varepsilon\omega\sin(\omega t)(S^2 - y^2) + 2(1 + \varepsilon\cos(\omega t)),$$

inducing small perturbation in time around a parabolic solution if $\varepsilon \ll 1$. Indeed, the exact solution for this equation is:

$$u(y,t) = (S^2 - y^2)f(t), \quad f(t) = (1 + \varepsilon\cos(\omega t)).$$

Consider a functional of the form:

$$j(S,t) = S^m u_y(y = S, t), \quad m \in \mathbb{N}^* \tag{8.17}$$

involving instantaneous state quantities. The sensitivity with respect to S is:

$$j_s(S,t) = mS^{m-1}u_y(S,t) + S^m u_{ys}(S,t). \tag{8.18}$$

The first term is the instantaneous incomplete sensitivity.

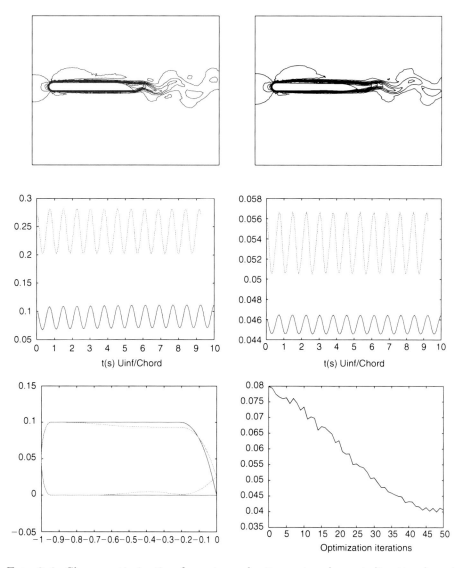

FIG. 8.4. Shape optimization for noise reduction using descent direction based on incomplete sensitivities. Upper: Iso-$|u|$ for the initial and optimized shapes. Middle: lift and drag evolution in time for the initial and optimized shapes. Lower: Initial and final shapes and evolution of J.

As we have:

$$u_y(S,t) = -2Sf(t) \quad \text{and} \quad u_{y_S}(S,t) = -2f(t),$$

one can express the different contribution in (8.18).

$$j_S(S,t) = mS^{m-1}(-2Sf(t)) + S^m(-2f(t)).$$

Comparing with $-2(m+1)S^m f(t)$, one sees that the approximation of the gradient based on this incomplete sensitivity is accurate and its precision increases with m. Most importantly, the incomplete sensitivity always has the right sign. It is obvious that the analysis still holds if the functional involves a time integral:

$$J(S,T) = \int_{(0,T)} S^m u_y(y=S,t)\, dt.$$

Now, if the functional involves an integral over the domain:

$$J(S,T) = \int_{(0,T)\times(-S,S)} j(y,t)\, dt\, dy,$$

one can still make the analysis above and see the importance of different contributions:

$$J_S(S,T) = \int_{(0,T)\times(-S,S)} (mS^{m-1}u_y(S,t) + S^m u_{yS}(y,t))\, dt\, dy$$

$$+ \int_{(0,T)} [S^m u_y(y,t)]_{\pm S}\, dt.$$

Again, an incomplete evaluation of the sensitivity is accurate because $u_{yS} = 0$. One also notices that if m is odd the last integral vanishes, even though this integral is cheap to evaluate as it does not involve any state sensitivity with respect to S. Anyway, incomplete sensitivity seems efficient but, as we said, it only holds for very special functionals.

Now, if j has no particular form the sensitivity reads:

$$J_S(S,T) = \int_{(0,T)\times(-S,S)} (j_S + j_u u_S)\, dt\, dy + \int_{(0,T)} j(\pm S,t)\, dt.$$

In this expression only u_S is costly to evaluate as it requires linearization of the state equation.

The solution of the linearized state equation (8.16) permits us to write for all functions v:

$$0 = \int_{(0,T)\times(-S,S)} ((u_S)_t - (u_S)_{yy} - F_S)v\, dt\, dy.$$

Integrating by part, it gives:

$$0 = \int_{(0,T)\times(-S,S)} u_S(-v_t - v_{yy})\, dt\, dy + \int_{(0,T)\times(-S,S)} -F_S v\, dt\, dy + \int_{(-S,S)} [vu_S]_T^0\, dy.$$

Again, let us introduce an adjoint problem. This is suitable when the dimension of the control space parameter is large but it is still instructive here.

$$v_t - v_{yy} = j_u, \qquad v(y,T) = v(\pm S, t) = 0. \tag{8.19}$$

Therefore, with v a solution of the backward adjoint equation (8.19) with the chosen boundary and final conditions one has:

$$\int_{(0,T)\times(-S,S)} u_s j_u \, dt \, dy = -\int_{(-S,S)} u_s(0) v(0) dy + \int_{(0,T)\times(-S,S)} F_s v \, dt \, dy,$$

where, unlike with the linearized equation, with S of any dimension v is computed only once before assembling the right-hand-side above.

Here, the state equation is linear and no storage of intermediate states is necessary in the adjoint calculation. We saw that if the state equation is nonlinear, solution of the adjoint equation requires the storage of all intermediate states between 0 and T. For instance, consider

$$u_t + u u_y - u_{yy} = F(t), \quad u(S,t) = u(-S,t) = 0.$$

For the same functional, the adjoint equation in this case reads

$$v_t + u v_y - v_{yy} = j_u, \qquad v(y,T) = v(\pm S, t) = 0,$$

where u is now present in the left-hand-side of the equation to be solved backward from T to 0.

8.4.2 Data mining and adjoint calculation

Let us give a more formal presentation for at what we said above and see how data compression techniques can be used to avoid the storage difficulty mentioned in adjoint calculations. Rewrite (8.13) as:

$$\min_{S \in O_{ad}} J(S, \mathcal{U}(S)) \tag{8.20}$$

and suppose the admissible space $O_{ad} \subset \mathbf{R}^n$ and $\mathcal{U}(S) = \{u(t, q(S)), t_i \in [0,T], i = 1, ..., K\}$ is an ensemble of K snapshots of the instantaneous vector states in $\mathbf{R}^N$. Usually, in applications we have $n \ll N, K$.

The direct simulation loop can be seen as:

$$J : S \to q(S) \to u(t, q(S)) \to \mathcal{U} \to J(S, \mathcal{U}). \tag{8.21}$$

The gradient of J with respect to S is given by:

$$J' = J_s(S) + J_q(S) \, q_s(S) + \mathcal{F}\left(J_\mathcal{U}(S,\mathcal{U}), \, \mathcal{U}_s(S,\mathcal{U})\right). \tag{8.22}$$

J' is a vector of size n and $\mathcal{U}$ is an ensemble of K state snapshots. $J_\mathcal{U}$ and $\mathcal{U}_s$ are defined as (denoting $u^i = u(t_i, S)$):

$$J_\mathcal{U} = (J_{u^1}, ..., J_{u^K}), \quad \mathcal{U}_s = (u_s^1, ..., u_s^K)$$

where J_{u^i} are vectors of size N and u_S^i ($N \times n$) matrices. $\mathcal{F}$ defines the interactions between different snapshots sensitivities. For instance, with a functional measuring a difference in time for some observation:

$$J = j(u^i) - j(u^j), \quad i = 2,..,K, \quad j = 1,..K-1, \quad i > j$$

$\mathcal{F}$ is given by:

$$\mathcal{F} = J_\mathcal{U}(S,\mathcal{U}) \, \Lambda \, \mathcal{U}_S(S,\mathcal{U})$$

with Λ a $(N \times n) \times K$ matrix with zero at all components, $+1$ for indices corresponding to u_S^i and -1 for those corresponding to u_S^j. If the state does not depend on time, the last product in (8.22) reduces to $J_u u_S$ which is a vector of size N.

As we saw in our model problem, as long as $\mathcal{U}$ is somehow available, the major part of the cost of this evaluation is due to $\mathcal{U}_S$ in the last term of (8.22). We also saw that in high-dimension optimization problems, one needs to have $\mathcal{U}$ available during backward adjoint integration. [9] Because of its complexity, this is not satisfactory for practical applications. It can be seen as solving a large linear system by assembling the matrix then inverting it which is highly inefficient. To overcome this difficulty, we would like to replace the loop (8.21) by an alternative low-complexity loop:

$$J : S \to q(S) \to u(t, q(S)) \to \tilde{\mathcal{U}} \to J(S, \mathcal{U}(\tilde{\mathcal{U}}))). \tag{8.23}$$

Data mining techniques (POD, PCA, SOM) mentioned in Chapter 7 permit us to build a low-dimension $\tilde{\mathcal{U}}$ to reduce storage and ease the manipulation of elements of $\mathcal{U}$. To be efficient, the incoming states should be assimilated by this model using some Kalman filtering strategy. Suppose one has k orthonormalized vectors $U_{i=1,...,k} \in \mathbf{R}^N$, $(U_i, U_j) = \delta_{i,j}, i,j = 1,...,k$, built from incoming u^j:

$$u^j = \sum_{i=1,...,k} (u^j, U_i) \, U_i + e^j, \quad j = 1,...,K, \quad k \ll K.$$

The errors e^j is small or new elements U_i should be introduced. Once this is established, it permits us to recover any u^j during backward time integration at a much lower storage cost. An upper bound of the required complexity can be obtained from signal sampling theory [10] [11] or from more efficient and recent compressive sampling theory [2] for data recovery from incomplete data.

---

[9] In Chapter 5 we mentioned that the checkpointing algorithm reduces the storage complexity from K states to $\log_2(K)$.

[10] The Shannon sampling theory states that exact reconstruction of a continuous time signal from its samples is possible if the signal is limited in band and if the sampling frequency is greater than twice the signal bandwidth. In other words, equidistributing data with at least two points per cycle of highest frequency allows exact reconstruction.

References

[1] Candes, E. Romberg, J. Tao, T. (2006). Robust uncertainty principles: Exact recovery from highly incomplete Fourier information, *IEEE Trans. Inform. Theory,* **52(2)**, 489-509.

[2] Candes, E. Romberg, J. and Tao, T. (2006). Stable signal recovery from incomplete and inaccurate measurements. *Comm. Pure Appl. Math,* **59(8)**, 1207-1223.

[3] Dadone, A. Mohammadi, B. and Petruzzelli, N. (1999). Incomplete sensitivities and BFGS methods for 3D aerodynamical shape design, *INRIA* **3633**.

[4] Gilbert, J.C. Le Vey, G. and Masse, J. (1991). La différentiation automatique des fonctions représentées par des programmes, *INRIA report* **1557**.

[5] Griewank, A. (1995), Achieving logarithmic growth of temporal and spatial complexity in reverse automatic differentiation, *Optim. Meth. Software,* **1**, 35-54.

[6] Marsden, A. Wang, M. and Koumoutsakos, P. (2002). *Optimal aeroacoustic shape design using approximation modelling,* CTR briefs 2002.

[7] Medic, G. B. Mohammadi, B. and S. Moreau, S. (1998). Optimal airfoil and blade design in compressible and incompressible flows, *American Institute of Aeronautics and Astronautics,* **98-2898**.

[8] Mohammadi, B. (1997). A new optimal shape design procedure for inviscid and viscous turbulent flows, *Int. J. Num. Meth. Fluid,* **25**, 183-203.

[9] Mohammadi, B. (1999). Flow control and shape optimization in aeroelastic configurations, *American Institute of Aeronautics and Astronautics,* **99-0182**.

[10] Rostaing, N. Dalmas, S. and Galligo, A. (1993). Automatic differentiation in Odyssee, *Tellus,* **45a(5)**, 558-568.

[11] Shannon, C. E. (1949). Communication in the presence of noise, *Proc. Institute of Radio Engineers,* **37:1**, 10-21, Jan. 1949. *Proc. IEEE,* **86(2)**, 1998.

9

CONSISTENT APPROXIMATIONS AND APPROXIMATE GRADIENTS

9.1 Introduction

In this chapter we shall give some rigor to the concepts discussed in the previous chapter and apply the theory of consistent approximations introduced by E. Polak [9] to study how the discretization parameters (e.g. the mesh) in an optimization problem can be refined within the algorithm loop: can we do some preliminary computations on a coarse grid, then use the solution as an initializer for a finer grid and so on ?

In the process we shall find the right spaces to fit the theory and understand the connection between the discrete and the continuous optimal shape design problems; we will discover which is the best norm and the best variable for shape design and in what sense the discrete gradients should converge to the continuous one. Application to academic test cases and to real life wing design will be given. This chapter summarizes our research done in cooperation with N. Dicesaré and E. Polak [3,8].

9.2 Generalities

We wish to merge grid refinement into the optimization loop so as to do most of the optimization iterations on a coarse mesh and a few on the fine meshes. This cannot be done at random, some theory is needed. To this end consider

$$\min_{z \in Z} J(z),$$

and its discretization (h could be the mesh size)

$$\min_{z \in Z_h} J_h(z),$$

where Z_h is a finite dimensional subspace of the Hilbert space Z.

Consider the method of steepest descent to achieve precision ϵ with the Armijo rule for the step size with parameters $0 < \alpha < \beta < 1$.

Steepest descent with Armijo's rule

while $\|\text{grad}_z J_h(z^m)\| > \epsilon$ **do**
{
$z^{m+1} = z^m - \rho \text{grad}_z J_h(z^m)$
where ρ is any number satisfying, with $w = \text{grad}_z J_h(z^m)$

$$\beta\rho\|w\|^2 < J_h(z^m - \rho w) - J_h(z^m) < -\alpha\rho\|w\|^2$$

$m := m+1;$
}

Now consider the same algorithm with grid refinement with parameter $\gamma > 1$.

Steepest descent with Armijo's rule and mesh refinement

while $h > h_{min}$ **do**
{
while $\|\text{grad}_z J_h(z^m)\| > \epsilon h^\gamma$ **do**
{
$z^{m+1} = z^m - \rho \text{grad}_z J_h(z^m)$
where ρ is any positive real such that, with $w = \text{grad}_z J_h(z^m)$

$$\beta\rho\|w\|^2 < J_h(z^m - \rho w) - J_h(z^m) < -\alpha\rho\|w\|^2$$

$m := m+1;$
}
$h := h/2;$
}

For this method the convergence is fairly obvious because at each grid level the convergence is that of the gradient method, and when the mesh size is decreased the norm of the gradient is required to decrease as well. Indeed, at least for the sequence m_j for which the test on the gradient fails, we have: $\text{grad}_z J_h(z^{m_j}) \to 0$ as $j \to \infty$. To prove the convergence of the entire sequence z^m is more difficult, even in strictly convex situations, because nothing ensures that J_h decreases when h is decreased.

Another possible gain in speed arises from the observation that we may not need to compute the exact gradient $\text{grad}_z J_h$! Indeed in many situations it involves solving PDEs and/or nonlinear equations by iterative schemes; can we stop the iterations earlier on coarse meshes? To this end assume that N is an iteration parameter and that $J_{h,N}$ and $\text{grad}_z J_{h,N}$ denote approximations of J_h and $\text{grad}_z J_h$ in the sense that

$$\lim_{N\to\infty} J_{h,N}(z) = J_h(z), \quad \lim_{N\to\infty} \text{grad}_{zN} J_{h,N}(z) = \text{grad}_z J_h(z).$$

Now consider the following algorithm with additional parameter K and $N(h)$ with $N(h) \to \infty$ when $h \to 0$:

Steepest descent with Armijo's rule, mesh refinement and approximate gradients

while $h > h_{\min}$
{
while $|\mathrm{grad}_{zN} J^m| > \epsilon h^\gamma$
{
try to find a step size ρ with $w = \mathrm{grad}_{zN} J(z^m)$

$$-\beta\rho\|w\|^2 < J(z^m - \rho w) - J(z^m) < -\alpha\rho\|w\|^2$$

if success **then**
$\{z^{m+1} = z^m - \rho\mathrm{grad}_{zN} J^m;\ m := m+1;\}$
else $N := N + K$;
}
$h := h/2;\ N := N(h)$;
}

The convergence could be established from the observation that Armijo's rule gives a bound on the step size:

$$-\beta\rho\mathrm{grad}_z J \cdot h < J(z+\rho h) - J(z) = \rho\mathrm{grad}_z J \cdot h + \frac{\rho^2}{2}J''hh$$

$$\Rightarrow \rho > 2(\beta-1)\frac{\mathrm{grad}_z J \cdot h}{J''(\xi)hh}$$

so that

$$J^{m+1} - J^m < -2\frac{\alpha(1-\beta)}{\|J''\|}|\mathrm{grad}_z J|^2.$$

Thus at each grid level the number of gradient iterations is bounded by $O(h^{-2\gamma})$. Therefore the algorithm does not jam and as before the norm of the gradient decreases with h.

A number of hypotheses are needed for the above, such as C^1 continuity with respect to parameters, bounded ness from below for J, etc. To make the above more precise we recall the hypothesis made by E. Polak in [9].

9.3 Consistent approximations

More precisely, consider an optimization problem with linear constraints

$$(\mathcal{P}) \quad \min_{z \in \mathcal{O}} J(z)$$

where $\mathcal{O}$, the set of admissible variables, is the subset of a Hilbert space $\mathcal{H}$ whose scalar product is denoted by $\langle .,. \rangle$.

Then consider a finite dimensional approximated problem

$$(\mathcal{P}_h) \qquad \min_{z_h \in \mathcal{O}_h} J_h(z_h).$$

Assume that $J : \mathcal{O} \longrightarrow R$ and $J_h : \mathcal{O}_h \longrightarrow R$ are continuous and differentiable, and that $\mathcal{O}_h \subset \mathcal{O}$. The *optimality functions* of the problems are

$$\theta(z) = -\|\mathbf{P}\operatorname{grad}_z J(z)\|, \quad \theta_h(z) = -\|\mathbf{P}_h \operatorname{grad}_z J_h(z)\|$$

where $\mathbf{P}$ and $\mathbf{P}_h$ are the projection operator from $\mathcal{H}$ to $\mathcal{O}$ and $\mathcal{H}$ to $\mathcal{O}_h$, respectively. Recall the definition of projectors, for instance

$$\langle \mathbf{P}z, w \rangle = \langle z, w \rangle \quad \forall w \in \mathcal{O}, \quad \mathbf{P}z \in \mathcal{O}.$$

Notice that θ (resp. θ_h) is always negative and if it is zero at $\tilde{z}$ then this point is a candidate optimizer for $\mathcal{P}$ (resp. $\mathcal{P}_h$).

We assume that both θ and θ_h are at least sequentially upper-semi-continuous in z.

9.3.1 Consistent approximation

For an *optimality function* θ_h for $\mathcal{P}_h$ with similar properties, the pairs $\{\mathcal{P}_h, \theta_h\}$ are *consistent approximations* to the pair $(\mathcal{P}, \theta)$, if

1. as $h \to 0$, for every $z \in \mathcal{O}$, there exist a sequence $\{z_h\}$ with $z_h \in \mathcal{O}_h$, such that $z_h \longrightarrow z$; for every $\tilde{z}$, there exist a sequence $\forall \{z_h\}_h$ such that $z_h \in \mathcal{O}_h$ and $z_h \longrightarrow \tilde{z}$, and we have: $\tilde{z} \in \mathcal{O}$ (*consistency*);
2. $J_h(z_h) \longrightarrow J(z)$, as $h \to 0$ (*continuity in h*);
3. $\forall \{z_h\}_h$ of $\mathcal{O}_h$ converges to $\tilde{z}$, $\overline{\lim_{h \to 0}} \theta_h(z_h) \le \theta(\tilde{z})$.

We summarize these hypotheses by the key-words, continuity in z and h for J and J_h, upper semi-continuity in z and h for θ and θ_h, and inclusion and consistency for $\mathcal{O}_h$. Notice that the first two items are the definition of epi-convergence of $\mathcal{P}_h$ to $\mathcal{P}$ [9].

9.3.2 Algorithm: conceptual

1. Choose a converging sequence of discretization spaces $\{\mathcal{O}_{h_n}\}$ with $\mathcal{O}_{h_n} \subset \mathcal{O}_{h_{n+1}}$ for all n. Choose $z^0, \epsilon^0, \beta \in]0,1[$.
2. Set $n = 0, \epsilon = \epsilon^0, h = h_0$.
3. Compute z_m^n by performing m iterations of a descent algorithm on $\mathcal{P}_h$ from the starting point z^n so as to achieve

$$\theta_h(z_m^n) > -\epsilon.$$

4. Set $\epsilon = \beta\epsilon, h = h_{n+1}, z^{n+1} = z_m^n, n = n+1$ and go to Step 3.

The following theorem is from [9].

9.3.2.1 Theorem
If $\mathcal{P}_h$ are consistent approximations of $\mathcal{P}$ then any accumulation point z^* of $\{z^n\}$ generated by Algorithm 9.3.2 satisfies $\theta(z^*) = 0$.

9.4 Application to a control problem

For a bounded domain Ω with boundary Γ consider the following model problem, which is in the class of problems that arise when a transpiration condition is used to simplify an optimal shape design problem:

$$\min_{v \in L^2(\Gamma)} \left\{ J(v) = \|u - u_d\|_1^2 \; : \; u - \Delta u = f \text{ in } \Omega, \; \frac{\partial u}{\partial n}\Big|_\Gamma = v \right\},$$

where

$$\|u - u_d\|_1^2 = \int_\Omega [(u - u_d)^2 + |\nabla(u - u_d)|^2].$$

Discretize by triangular finite elements with $u \in V_h$ continuous piecewise linear on a triangulation $\mathcal{T}_h$:

$$\min_{v \in L_h} \left\{ J_h(v) = \|u - u_{dh}\|_1^2 \; : \; \int_\Omega (uw + \nabla u \nabla w) = \int_\Gamma vw \right\},$$

where L_h is the space of piecewise constant functions on Γ_h and where u_{dh} is an approximation of u_d in V_h (the interpolation of u_d, for instance).

The variations of J with respect to variations in v are computed as follows (all equalities are up to higher order terms in δv):

$$\delta J = 2 \int_\Omega [(u - u_d)\delta u + \nabla(u - u_d) \cdot \nabla \delta u]$$

with

$$\int_\Omega (\delta u w + \nabla \delta u \nabla w) = \int_\Gamma \delta v w.$$

Therefore the "derivative" of J with respect to v is $J' = 2(u - u_d)$ because

$$\delta J = 2 \int_\Gamma \delta v (u - u_d).$$

The same computation is performed for J_h without change and leads to

$$\delta J_h = 2 \int_\Gamma \delta v_h (u_h - u_{dh}).$$

Let us define

$$\theta = -\|u - u_d\|_{0,\Gamma}, \quad \theta_h = - \sup_{v_h \in L_h: \|v_h\|_0 = 1} \int_\Gamma v_h (u_h - u_{dh}).$$

Note that if s_j is a segment of length $|s_j|$ of the discretization of Γ, θ_h is the L^2 norm of the piecewise constant function:

$$\Theta_h = \frac{1}{|s_j|} \int_{s_j} (u_h - u_{dh}).$$

With the method of steepest descent, Algorithm 9.3.2 becomes:

9.4.1 Algorithm: control with mesh refinement

Assume that we have an automatic mesh generator which yields a triangulation of Ω in a deterministic manner once the discretization of the boundary Γ is chosen.

1. Choose an initial segmentation h_0 of Γ. Set $v_{h_0,0}^0 = 0$, $\epsilon^0 = 1$, $\beta = 0.5$ (for instance).
2. Set $n = 0$, $\epsilon = \epsilon^0$, $h = h_0$.
3. Compute $v_{h,m}^n$ by performing m iterations of the type

$$v_{h,m+1}^n = v_{h,m}^n - \rho(u_{h,m}^n - u_{dh})$$

where ρ is the best real number which minimizes J_h in the direction $u_{h,m}^n - u_{dh}$ and $u_{h,m}^n$ is the FEM solution of the PDE with $v_{h,m}^n$, until

$$\left(\sum_j \left(\frac{1}{|s_j|} \int_{s_j} (u_{h,m}^n - u_{dh}) \right)^2 |s_j| \right)^{1/2} < \epsilon.$$

4. Divide some segments on Γ. Set $\epsilon = \epsilon/2$, $v_{h,0}^{n+1} = v_{h,m}^n$, $n = n+1$ and go to step 3.

9.4.2 Verification of the hypothesis

- *Inclusion* When a segment of the discrete boundary is divided into two segment we do not have exactly

$$h' < h \Rightarrow L_{h'} \subset L_h$$

because the boundary is curved; but this is a minor technical point and we can consider that L_h refers to the curved boundary divided into segments and use an over parametric triangular finite element for the triangles which use these edges.

- *Continuity* We have the following implications

$$v^n \xrightarrow{L^2(\Gamma)} v, \Rightarrow u^n \xrightarrow{H^1(\Omega)} u, \Rightarrow J(v^n) \to J(v)$$
$$\Rightarrow \|u^n - u_d\|^2 \to \|u - u_d\|^2.$$

So J, θ, J_h and θ_h are continuous with respect to v because

$$v_j^n \xrightarrow{\mathcal{R}} v_j, \Rightarrow u_h^n \xrightarrow{H^1(\Omega)} u_h,$$

$$\Rightarrow \sup_{v_h \in L_h:\ \|v_h\|=1} \int_\Gamma v_h(u_h^n - u_{dh}) \to \sup_{v_h \in L_h:\ \|v_h\|=1} \int_\Gamma v_h(u_h - u_{dh}).$$

- *Consistency* is obvious by simple discretization of Γ

$$\forall v \in L^2(\Gamma),\ \exists\ \{v_h\},\ v_h \to v\ :\ u_h \to u,\ \|u_h - u_{dh}\|^2 \to \|u - u_d\|^2.$$

and also because if $v_h \in L_h \to v$ then $v \in L^2(\Gamma)$.
- *Continuity* as $h \to 0$ of J_h and of θ_h. This is usually the most stringent property to verify but here it is easy by the convergence properties of the finite element method:

$$v_h \to v,\ \Rightarrow u_h \to u,\ \Rightarrow J_h(u_h) \to J(u),$$

$$\Rightarrow \sup_{v_h \in L_h:\ \|v_h\|=1} \int_\Gamma v_h(u_h - u_{dh}) \to \sup_{v \in L^2:\ \|v\|=1} \int_\Gamma v(u - u_d).$$

For simplicity we have purposely chosen an example which does not require an adjoint, but for those which do there is hardly any additional difficulty, for instance:

$$J(v) = \|u - u_d\|_{0,\Omega}^2 \Rightarrow J' = p|_\Gamma\ :\ p - \Delta p = 2(u - u_d),\ \frac{\partial p}{\partial n} = 0$$

$$v^n \xrightarrow{L^2(\Gamma)} v,\ \Rightarrow u^n \xrightarrow{H^1(\Omega)} u,\ \Rightarrow p^n \xrightarrow{H^1(\Omega)} p \Rightarrow J'^n \to J'.$$

9.4.3 Numerical example

Consider an ellipse with a circular hole and boundary control on the hole. Table 9.1 shows the performance of the method.

Table 9.1 History of the convergence and values of the cost function

Mesh\Iteration	Iter1	Iter2	Iter3
Mesh1	3818.53	1702.66	1218.47
Mesh2	469.488	285.803	
Mesh3	113.584	74.4728	
Mesh4	23.1994		
Mesh5	7.86718		

Figure 9.1 show the triangulations generated each time the boundary is refined and table 9.1 shows the values of the criteria for each iteration and each mesh. Algorithm 9.4.1 has performed three iterations on mesh 1, then went to mesh 2 because θ_h has decreased below the threshold ϵ; two iterations were sufficient on mesh 2, etc. The computing time saved is enormous when compared to the same done on the finest mesh only.

9.5 Application to optimal shape design

Consider a simple model problem where the shape is to be found that brings u, solution of a PDE, nearest to u_d in a subregion D of the entire domain Ω. The unknown shape Γ is a portion of the entire boundary $\partial\Omega$: it is parameterized by its distance α to a reference smooth boundary Σ.

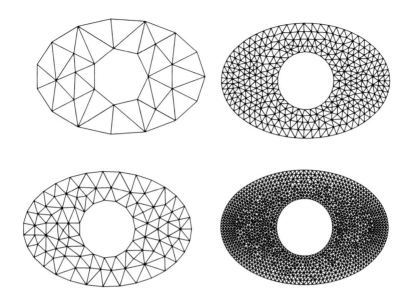

FIG. 9.1. *The first four meshes.*

9.5.1 *Problem statement*

More concretely given,

$$D \subset \Omega, \ u_d \in H^1(D), \ g \in H^1(\Omega), \ I \subset K \subset \mathcal{R}, \ \Sigma = \{x(s) \ : \ s \in K\}$$

we consider

$$\min_{\alpha \in H_0^2(I)} J(\alpha) = \int_D (u - u_d)^2 \quad \text{subject to}$$

$$u - \Delta u = 0 \ \text{in} \ \Omega(\alpha), \quad \frac{\partial u}{\partial n}|_{\Gamma(\alpha)} = g,$$

$$\text{where} \ \Gamma(\alpha) \equiv \partial \Omega(\alpha) = \{x(s) + \tilde{\alpha}(s) n(x(s)) \ : \ s \in K\}$$

and where $\tilde{\alpha}$ is the extension by zero in K of α.

Recall that

$$H_0^2(I) = \{\alpha \in L^2(I) \ : \ \alpha', \alpha'' \in L^2(I), \ \alpha(a) = \alpha'(a) = 0 \ \forall a \in \partial I\}$$

and that $\|\alpha''\|_0 = \|d^2\alpha/ds^2\|_0$ is a norm in that space.

The problem may not have a unique solution. To insure uniqueness one may add a regularizing term and replace the cost function by

$$J_\epsilon(\alpha) = \int_D (u - u_d)^2 + \epsilon \int_\Sigma \left|\frac{d^2\alpha}{ds^2}\right|^2.$$

Let us denote the unknown part of the boundary by
$$S(\alpha) = \{x(s) + \alpha(s)n(x(s)) \; : \; s \in I\}.$$
For simplicity, let us assume that g is always zero on S.

9.5.2 Discretization

The discrete problem is

$$\min_{\alpha \in L_h \subset H_0^2(I)} J_\epsilon(\alpha) = \int_D (u - u_d)^2 + \epsilon \int_\Sigma |\frac{d^2\alpha}{ds^2}|^2$$

$$\text{subject to} \qquad \int_{\Omega(\alpha)} (uv + \nabla u \nabla v) = \int_{\Gamma(\alpha)} gv$$

where V_h is the usual Lagrange finite element space of degree 1 on triangles except that the boundary triangles have a curved side because in the following definition $S(\alpha)$ is a cubic spline.

The space L_h is the finite dimensional subspace of $H_0^2(I)$ defined as the set of cubic splines which passes through the vertices which we would have used otherwise to define a feasible polygonal approximation of the boundary. This means that the discretization of Ω is done as follows:

1. Given a set of n_f boundary vertices $q^{i_1}, ..., q^{i_{n_f}}$, construct a polygonal boundary near Σ.
2. Construct a triangulation of the domain inside this boundary with an automatic mesh generator, i.e. mathematically the inner nodes are assumed to be linked to the outer ones by a given map
$$q^j = Q^j(q^{i_1}, ..., q^{i_{n_f}}), \quad n_f < j < n_v.$$
3. Construct $\Gamma(\alpha)$, the cubic splines from $q^{i_1}, ..., q^{i_{n_f}}$, set α to be the normal distance from Σ to $\Gamma(\alpha)$.
4. Construct V_h by using triangular finite elements and overparametric curved triangular elements on the boundary.

This may seem complex but it is a handy construction for several reasons. Among other things, the discrete regularized cost function J_h coincides with the regularized continuous J and L_h is a finite subspace of the (infinite) set of admissible parameters H_0^2.

9.5.3 Optimality conditions: the continuous case

As before, by calculus of variations

$$\delta J = 2 \int_D (u - u_d)\delta u + 2\epsilon \int_\Sigma \frac{d^2\alpha}{ds^2} \frac{d^2\delta\alpha}{ds^2},$$

with $\delta u \in H^1(\Omega(\alpha))$ and

$$\int_{\Omega(\alpha)} (\delta uv + \nabla \delta u \nabla v) + \int_\Sigma \delta\alpha(uv + \nabla u \nabla v) = 0.$$

Introduce an adjoint $p \in H^1(\Omega(\alpha))$:

$$\int_{\Omega(\alpha)} (pq + \nabla p \nabla q) = 2 \int_D (u - u_d) q, \quad \forall q \in H^1(\Omega(\alpha)),$$

i.e.

$$p - \Delta p = I_D u, \quad \frac{\partial p}{\partial n} = 0.$$

Then

$$\delta J = -\int_\Sigma \delta \alpha \left(up + \nabla u \nabla p - 2\epsilon \frac{d^4 \alpha}{ds^4} \right).$$

Definition of θ We should take

$$\theta = -\|up + \nabla u \nabla p - 2\epsilon \frac{d^4 \alpha}{ds^4}\|_{-2},$$

i.e. solve

$$\frac{d^4 \Theta}{ds^4} = up + \nabla u \nabla p \quad \text{on } I, \quad \Theta = \frac{d\Theta}{ds} = 0 \quad \text{on } \partial I$$

and take $\theta = -\|\Theta'' + 2\epsilon \alpha''\|_{0,\Sigma}$.

9.5.4 Optimality conditions: the discrete case

Let w^j be the hat function attached to vertex q^j. If some vertices q^j vary by δq_j we define

$$\delta q_h(x) = \sum_j \delta q_j w^j(x)$$

and we know that [7]

$$\delta w^k = -\nabla w^k \cdot \delta q_h, \quad \int_{\delta \Omega} f = \int_\Omega \nabla \cdot (f \delta q_h) + o(|\delta q_h|).$$

Hence

$$J(\alpha + \delta \alpha) = 2 \int_D (u_h - u_{dh}) \delta u_h + 2\epsilon \int_\Sigma \frac{d^2 \alpha}{ds^2} \frac{d^2 \delta \alpha}{ds^2}.$$

Furthermore, and by definition of δu_h

$$\delta \sum_i u_i w^i = \sum_i (\delta u_i w^i + u_i \delta w^i) = \delta u_h + \delta q_h \cdot \nabla u_h$$

the partial variation δu_h is found by

$$\delta \int_{\Omega(\alpha)} (u_h w^j + \nabla u_h \nabla w^j) = \int_\Omega (\nabla \cdot (uw^j \delta q_h) + \delta u_h w^j + \nabla \delta u_h \nabla w^j)$$

$$+ \int_\Omega (u_h \delta q_h \cdot \nabla w^j + \nabla u_h \nabla \delta q_h \nabla w^j + u_h \delta w^j + \nabla u_h \nabla \delta w^j) = 0.$$

Hence,

$$\int_\Omega (\delta u_h w^j + \nabla \delta u_h \nabla w^j)$$
$$= \int_\Omega (\nabla u_h (\nabla \delta q_h + \nabla \delta q_h^T) \nabla w^j - (u_h w^j + \nabla u_h \cdot \nabla w^j) \nabla \cdot \delta q_h).$$

So introduce an adjoint $p_h \in V_h$

$$\int_\Omega (p_h w^j + \nabla p_h \nabla w^j) = 2 \int_D (u_h - u_{dh}) w^j \quad \forall j.$$

And finally

$$\delta J_h = \int_\Omega (\nabla u_h (\nabla \delta q_h + \nabla \delta q_h^T) \nabla p_h$$
$$-(u_h p_h + \nabla u_h \cdot \nabla p_h) \nabla \cdot \delta q_h) + 2\epsilon \int_\Sigma \frac{d^2\alpha}{ds^2} \frac{d^2\delta\alpha}{ds^2}.$$

9.5.5 Definition of θ_h

Let $e^1 = (1,0)^T$, $e^2 = (0,1)^T$ be the coordinate vectors of R^2, and let χ^j be the vector of R^2 of components

$$\chi_k^j = \int_\Omega (\nabla u_h (\nabla w^j e^k + (\nabla w^j e^k)^T) \nabla p_h - (u_h p_u + \nabla u_h \cdot \nabla p_h) \nabla \cdot w^j e^k.$$

Because the inner vertices are linked to the boundary ones by the maps Q^j, let us introduce

$$\xi_k^j = \chi_k^j + \sum_{q^i \notin \Gamma} \chi^i \partial_{q_k^j} Q^i.$$

Then obviously

$$\delta J = \sum_1^{n_f} \xi^j \cdot \delta q^j + 2\epsilon \int_\Sigma \frac{d^2\alpha}{ds^2} \frac{d^2\delta\alpha}{ds^2}.$$

It is possible to find a β so as to express the first discrete sum as an integral on Σ of $\frac{d^2\beta}{ds^2} \frac{d^2\delta\alpha}{ds^2}$; it is a sort of variational problem in L_h:

$$\text{Find } \beta \in L_h \text{ such that } \int_\Sigma \frac{d^2\beta}{ds^2} \frac{d^2\lambda^j}{ds^2} = \xi^j \cdot n_\Sigma, \quad \forall j = 1, ..., n_f,$$

where λ^j is the cubic spline obtained by a unit normal variation of the boundary vertex q^j only.
Then the "derivative" of J_h is the function $\Theta_h : s \in I \to \beta''(s) + 2\epsilon\alpha''(s)$ and the function θ_h is

$$\theta_h = -\|\beta'' + 2\epsilon\alpha''\|_{L_0^2(I)}.$$

9.5.6 Implementation trick

This may be unnecessarily complicated in practice. A pragmatic summary of the above is that β is solution of a fourth-order problem and that its second derivative is the gradient, so why not set a discrete fourth-order problem on the normal component of the vertex itself. In the case $\epsilon = 0$ this would be

$$\frac{1}{h^4}[q'^{j+2}_n - 4q'^{j+1}_n + 6q'^j_n - 4q'^{j-1}_n + q'^{j-2}_n] = \xi^j,$$

$$q'^0_n = q'^1_n = q'^{n_f-1}_n = q'^{n_f}_n = 0$$

and then the norm of the second derivative of the result for θ_h:

$$\Theta_h \approx -\left(\sum_j \frac{1}{h^2}[q^{j+1}_n - 2q^j_n + q^{j-1}_n]^2\right)^{\frac{1}{2}}.$$

Another way is to notice that instead of a smoother on q^j we can construct displacements which have the same effects: Given ξ let

$$-\Delta \vec{u} - \nabla(\nabla \cdot \vec{u}) = 0, \quad u|_\Sigma = \xi$$

and then let

$$-\Delta \vec{v} = \vec{u}, \quad \frac{\partial \vec{v}}{\partial n} = 0.$$

Obviously

$$\xi \in H^s(\Sigma) \Rightarrow \vec{u} \in H^{s+1/2}(\Omega) \Rightarrow \vec{v}|_\Sigma \in H^{s+2}(\Sigma).$$

So $\Theta = v|_\Sigma$ is a smoothed version of ξ to the right degree of smoothing. The discretization of this method is obvious and it has another advantage in that Θ being defined everywhere all mesh points can be moved in the direction Θ.

9.5.7 Algorithm: OSD with mesh refinement

An adaptation of Algorithm 9.2 to this case is:

1. Choose an initial set of boundary vertices.
2. Construct a finite element mesh, and construct the spline of the boundary.
3. Solve the discrete PDE and the discrete adjoint PDE.
4. Compute Θ_h (or its approximation (cf. the subsection above)).
5. If $\theta_h = -\|\Theta_h\| > -\epsilon$ add points to the boundary mesh, update the parameters and go back to step 2.

9.5.8 Orientation

What the theory of consistent approximation has done for us is to:
- Give us an indication about the right spaces for the parameters (smooth radius of curvatures because of H_0^2 and fixed slopes at fixed end points).
- Give us a smoother before updating the mesh (a fourth-order smoother).
- Give us an error indicator to refine the mesh.

We need to verify all the hypotheses listed at the beginning of this chapter to make sure that Algorithm 9.5.7 converges (see [3] for the details of the proof). It is based on a result which is interesting in itself because it shows that the motion of the inner vertices can be neglected when compared to the motion of the boundary nodes.

9.5.8.1 Lemma

$$\delta J_h = -\int_\Sigma \left(\delta q_h \cdot n_\Sigma (u_h p_h + \nabla u_h \nabla p_h) + 2\epsilon \frac{d^4 \alpha_h}{ds^4} \right) + O(h \delta q_h) + o(h).$$

Proof We note that a change of variable $x = Q(X)$ in the following integral gives

$$\int_{Q^{-1}(\Omega)} ((u_h p_h + \nabla u_h (\nabla Q \nabla Q^T) \nabla p_h) \det \nabla Q^{-1} = \int_\Omega (u_h p_h + \nabla u_h \nabla p_h).$$

Take $X(x) = x + \delta q_h(x)$, then $\nabla Q = I - \nabla \delta q_h$, $\det \nabla Q^{-1} \approx 1 + \nabla \cdot \delta q_h$, so

$$\delta J_h = \int_\Omega (\nabla u_h (\nabla \delta q_h + \nabla \delta q_h^T) \nabla p_h - (u_h p_h + \nabla u_h \cdot \nabla p_h) \nabla \cdot q_h + ...$$

$$= \sum_k \int_{Q^{-1}(T_k) \setminus T_k} (u_h p_h + \nabla u_h \nabla p_h) + ...$$

$$= -\int_\Sigma \delta q_h \cdot n_\Sigma (u_h p_h + \nabla u_h \nabla p_h)$$

$$- \sum_k \int_{\partial T_k - \Sigma} \delta q_h \cdot \vec{n} (\nabla u_h \cdot \nabla p_h) + o(h) + ...$$

The last integral involves jumps of $\nabla u_h \cdot \nabla p_h$ across the edges of the elements; error estimates for the finite element method shows that this is $O(h)$ because in the limit $h \to 0$ these jumps are zero.

9.5.9 Numerical example

Effect of the smoother Consider first an inverse problem with incompressible inviscid flow modeled by

$$-\Delta \psi = 0 \text{ in } \Omega \qquad \psi_{in} = \psi_{out} = y \quad \psi_{y=0} = 0, \quad \psi_{top} = 1. \tag{9.1}$$

When Ω is a rectangular pipe $\Omega^* = (0,L) \times (0,1)$ then $\psi = y$ and the velocity u_d is uniformly equal to (1,0). Let $v_d = (0,1)i$, D be $(0,L) \times (0,1/2)$ and consider

$$\min_{\Omega} \int_D |\nabla \psi - v_d|^2 \text{ subject to (9.1) in } \Omega.$$

Assume that only the part of the top boundary Σ in the interval $x \in (1/3, 2/3)$ varies. By starting the optimal shape algorithm with a non-rectangular pipe we should recover the rectangular one. We apply the method without mesh refinement but with different scalar products to define the gradients. In one case we use the above theory, i.e. the scalar product of $H_0^2(\Sigma)$. In the second case we use $L^2(\Sigma)$, which amounts to directly using the variations of criteria without smoother. The results shown in Fig. 9.2 prove the necessity of the smoother; after four iterations the top boundary oscillates too much and the mesh is no longer compatible. In this case the mesh nodes are moved proportionally to the motion of the top boundary.

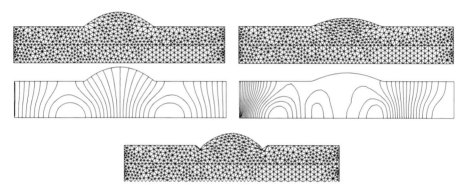

FIG. 9.2. Effect of the smoother (i.e. a good choice for the spaces). Initial guess (left) and after 10 iterations. Without the smoother after four iterations the mesh is no longer compatible (bottom). The level lines for the initial guess and the final results are shown; while the first ones are in the interval (0,0.1), the final ones are in (0,0.001). The exact solution is the straight pipe.

9.5.10 A nozzle optimization

Statement of the problem Consider a symmetric nozzle of length $L > 0$. Incompressible potential flow is modeled by

$$\varepsilon\varphi - \Delta\varphi = 0 \text{ in } \Omega, \quad \left.\frac{\partial\varphi}{\partial n}\right|_\Gamma = g|_\Gamma, \qquad (9.2)$$

where $0 < \varepsilon \ll 1$ is a regularizing constant used to avoid the singularity of Neumann conditions. The inflow and outflow velocities can be rescaled so that

$$g|_{\Gamma^1} = -1, \quad g|_{\Gamma^2} = \frac{|\Gamma^1|}{|\Gamma^2|}, \quad g|_{\Gamma^3 \cup \Gamma^4} = 0, \tag{9.3}$$

where $|\Gamma|$ denotes the length of Γ. We consider the inverse problem of designing a nozzle that gives a flow as close as possible to a prescribed flow u_d in a subset of Ω, say in a given region $D = [0, L] \times [0, d]$. One way to fulfill this requirement is through the optimization problem

$$\min_{\Gamma^3 \in \mathcal{O}'} J_0(\Gamma^3) = \int_D \|\nabla \varphi - u_d\|^2, \tag{9.4}$$

where $\mathcal{O}'$ the set of admissible boundaries:

$$\mathcal{O}' = \{\Gamma^3 : D \cap \Gamma^3 = \emptyset, \, \Omega \subset \Omega_0, \partial \Gamma^3 \subset \partial \Gamma^4\}, \tag{9.5}$$

for some given security rectangle $\Omega_0 = (0, L) \times (0, H)$. The upper parts of Γ^4 are horizontal.

As stated, the set $\mathcal{O}'$ is too large and the problem may not have a solution. Let $I = (a, b)$ be an interval and let $\mathcal{O}$ be the set of admissible boundaries which have a representation $\{x, y(x)\}_{x \in \bar{I}}$ in the (x, y) plane:

$$\mathcal{O} = \{\Gamma^3_\alpha = \{x, \alpha(x)\}_{x \in \bar{I}}, \, \alpha \in H^2(I) : d \leq \alpha(x) \leq H, \forall x \in I,$$
$$\alpha(a) = |\Gamma^1|, \alpha(b) = |\Gamma^2|, \alpha'(a) = \alpha'(b) = 0\}.$$

Note that $\mathcal{O}$ is a subspace of $H_0^2(I)$ and hence it inherits the Hilbert structure of $H_0^2(I)$. We will minimize the function (9.6). So our problem is

$$(\mathcal{P}) \quad \min_{\Gamma^3_\alpha \in \mathcal{O}} J(\Gamma^3_\alpha) = \int_D |\nabla \varphi - u_d|^2 + \varepsilon \int_I \left|\frac{d^2 \alpha}{dx^2}\right|^2 dx, \tag{9.6}$$

where φ is the solution of (9.2) and the second term is here to insure the existence of a solution to the optimization problem.

9.5.10.1 *Computation of the Gâteaux derivative* In one dimension, $H^s \hookrightarrow C^k$, $s > k + 1/2$, hence all admissible α are C^1 at least. We will use a norm on α associated with the scalar product of $H_0^2(I)$:

$$\langle \alpha, \beta \rangle_{H_0^2(I)} = \int_I \frac{d^2 \alpha}{dx^2} \frac{d^2 \beta}{dx^2} dx, \quad \|\alpha\|_{H_0^2(I)} = \langle \alpha, \alpha \rangle_{H_0^2(I)}^{\frac{1}{2}}. \tag{9.7}$$

Now, $\mathrm{grad}_\alpha J$, the $H_0^2(I)$ gradient of J, is related to the Gâteaux derivative $J'_\alpha(\beta)$ by

$$J'_\alpha(\beta) = \frac{d}{d\lambda} J(\Gamma^3_{\alpha + \lambda \beta})\big|_{\lambda=0} = \langle \mathrm{grad}_\alpha J, \beta \rangle_{H_0^2(I)} = \int_I \frac{d^4}{dx^4} \mathrm{grad}_\alpha J \cdot \beta$$

$$= \int_I J'_\alpha \cdot \beta.$$

Hence $\operatorname{grad}_\alpha J$ is the solution of

$$\frac{d^4}{dx^4}\operatorname{grad}_\alpha J = J'_\alpha \text{ in } I \quad \frac{d}{dx}\operatorname{grad}_\alpha J = 0 \text{ at } x = a \text{ and } x = b, \quad (9.8)$$

where J'_α is defined by the following theorem.
This scalar product will force the regularity of the boundaries when an iterative process like the gradient method is to be applied (see Fig. 9.2).

9.5.11 Theorem
The function J'_α is given by

$$J'_\alpha = -\frac{1}{n_2(\alpha)}\left(\varepsilon\varphi p + \nabla\varphi.\nabla p\right) + 2\varepsilon\frac{d^4\alpha}{dx^4}, \quad (9.9)$$

where $n_2(\alpha)$ is the y component of the exterior normal to Γ^3_α and $p \in H^1(\Omega)$ is the adjoint state, which is the solution of the Neumann problem

$$\int_\Omega (\varepsilon p w + \nabla p.\nabla w) = 2\int_D \nabla w.(\nabla \varphi - u_d)\,dxdy \quad \forall w \in H^1(\Omega). \quad (9.10)$$

Proof Let

$$\varphi'_\alpha(\beta) = \frac{d}{d\lambda}\varphi(\Gamma^3_{\alpha+\lambda\beta})|_{\lambda=0} \in H^1(\Omega), \quad (9.11)$$

the Gâteaux derivative of φ. It is the solution of

$$\forall w \in H^1(\Omega), \int_\Omega \left(\varepsilon\varphi'_\alpha(\beta)w + \nabla\varphi'_\alpha(\beta).\nabla w\right)dxdy \quad (9.12)$$

$$= -\int_{\Gamma_\alpha} \beta\left(\varepsilon\varphi w + \nabla\varphi.\nabla w - \frac{\partial}{\partial n}(gw) + \frac{gw}{R}\right)d\gamma \quad (9.13)$$

where R is the radius of curvature of Γ_α and $d\gamma$ the unit length element on the boundary of Γ_α. But in our case $g \equiv 0$ near and on Γ^3_α and $\beta = 0$ on $\Gamma^i_\alpha, i \neq 3$.
Then the variation of J is written as follows

$$\frac{d}{d\lambda}J(\Gamma^3_{\alpha+\lambda\beta})|_{\lambda=0} = 2\int_D \nabla\varphi'_\alpha(\beta).(\nabla\varphi - u_d)\,dxdy + 2\int_I \varepsilon\frac{d^2\alpha}{dx^2}\frac{d^2\beta}{dx^2}\,dx. \quad (9.14)$$

Next, introduce p as above and replace w by p in (9.10). As $\beta \in C^1$, we can use the mean value theorem for integrals and approximate the element of volume $dxdy$ by $\beta d\gamma$ thus using the definition of p with w replaced by $\varphi'_\alpha(\beta)$ and (9.12):

$$\frac{d}{d\lambda}J(\Gamma^3_{\alpha+\lambda\beta})|_{\lambda=0}$$

$$= -\int_{\Gamma^3_\alpha} \beta(\varepsilon\varphi p + \nabla\varphi.\nabla p)\,d\gamma + 2\int_I \varepsilon\frac{d^2\alpha}{dx^2}\frac{d^2\beta}{dx^2}$$

$$= \int_I \beta\left[2\varepsilon\frac{d^4\alpha}{dx^4} - \frac{1}{n_2(\alpha)}(\varepsilon\varphi p + \nabla\varphi.\nabla p)\right].$$

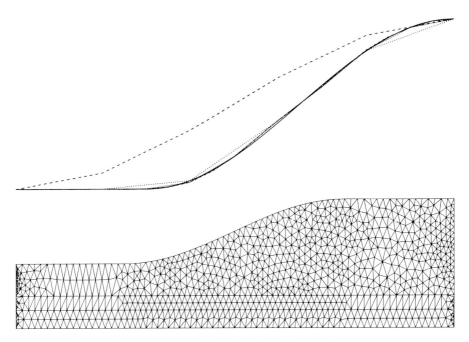

FIG. 9.3. Some shapes generated by the algorithm on its way to convergence and the triangulation for one of them.

9.5.12 Numerical results

We use a conjugate gradient in H_0^2 with mesh refinement (see Fig. 9.3). The initial mesh has 72 nodes, 110 triangles and five control points. In the end, after four refinements, the mesh is made of 9702 nodes, 18 980 triangles and 80 control points. We also compare with the results of the direct computation: we optimize directly on the big mesh (9702 nodes). It appears that computation with refinement is better (Fig. 9.4) but also faster than the direct computation (direct computation: 34h 11m 57s, iterative computation: 3h 37m 31s).

9.5.13 Drag reduction for an airfoil with mesh adaptation

This is a drag reduction problem for Euler flow with constraints on the lift and the volume treated by penalty. The cost function is given by:

$$J(x) = \frac{C_d}{C_d^0} + 100 \left| \frac{C_l - C_l^0}{C_l^0} \right| + 0.1 \left| \frac{\text{Vol} - \text{Vol}^0}{\text{Vol}^0} \right|,$$

where C_d and C_l are the drag and lift coefficients and Vol is the volume, superscript 0 denotes initial quantities. The initial airfoil is a RAE2822. The design takes place at Mach number 0.734 and Reynolds number 6.510^6 (Fig. 9.5, 9.6 and 9.7). The different ingredients are those described in Chapter 5. A Delau-

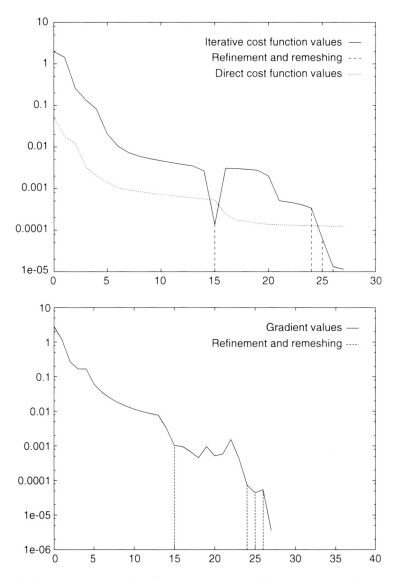

FIG. 9.4. Upper: cost function J versus iterations (direct computation and computation with refinement). Lower: norm of the gradient versus iterations.

nay mesh generator with adaptation by auxiliary metric techniques is used. The metric is based on the intersection of the metrics for the conservation variables.

With P1 discretization, fourth-order elliptic problems need mixed formulations which are not convenient for general 3D surfaces. Therefore, we use a local second-order elliptic smoother and, to avoid global smoothing of the deforma-

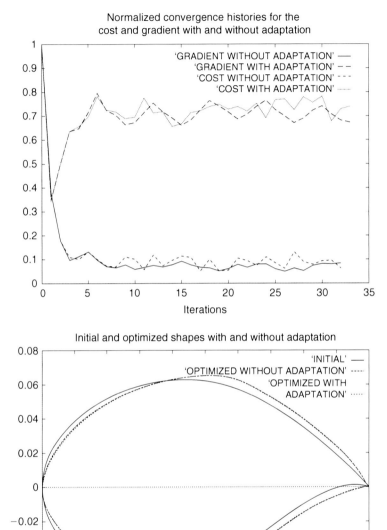

FIG. 9.5. Shape optimization in the transonic viscous turbulent regime: cost function and norm of gradients for airfoil optimization for the optimization alone and when combined with adaptation (top) and the initial and final shapes for each approaches (below). With adaptation, we can be confident that the mesh is always adequate for intermediate states.

tions, the smoothing is only applied locally to recover the behavior of fourth-order operators. This can be seen therefore as an alternative to the fourth order smoother (see Chapter 5 for more details on this smoother).

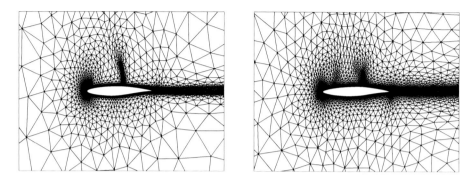

FIG. 9.6. Shape optimization in the transonic viscous turbulent regime. Initial (left) and final (right) meshes when using adaptation during optimization.

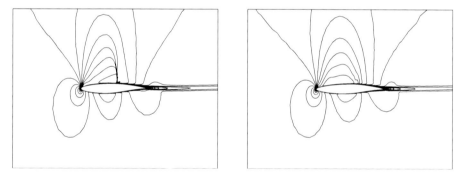

FIG. 9.7. Shape optimization in the transonic viscous turbulent regime. Initial (left) and final (right) iso-Mach contours for the previous meshes.

9.6 Approximate gradients

In control theory there is a practical rule that it is always better to use the exact gradient of the discrete problem rather than a discretized version of the gradient of the continuous problem. But in some cases it is difficult or unnecessarily expensive to compute the exact gradient of the discrete problem. For example, we have seen that in the cost function the variations due to the motions of the inner nodes are an order of magnitude smaller than those due to the boundary nodes. It would seem reasonable then to forget then. However, if this is done too soon the algorithm may jam. For nonlinear problems there would be a great speed-up if we could link the number of iterations of the nonlinear solver with the optimization loop index.

Another example comes from parallel computing. When the Schwarz algorithm is used to solves PDEs it is not easy to compute the derivatives of the discretized problems. We analyze one such problem below in the simpler context of boundary control.

9.6.1 A control problem with domain decomposition

Consider the boundary control problem (S is a given subset of the boundary Γ of an open bounded subset of R^d, Ω):

$$\min_{v \in L^2(S)} \{J(v) = \int_\Omega [(u - u_d)^2 + |\nabla(u - u_d)|^2] : \text{subject to} $$
$$u - \Delta u = 0 \text{ in } \Omega, \quad \frac{\partial u}{\partial n}|_S = \xi v \quad u_{\Gamma - S} = u_d\}. \tag{9.15}$$

The v-derivative of J is easy to obtain from

$$\delta J = 2 \int_\Omega ((u - u_d)\delta u + \nabla(u - u_d) \cdot \nabla \delta u + o(|v|)) = \int_S \xi(u - u_d)\delta v.$$

This is because the PDE in variational form is

$$\int_\Omega (uw + \nabla u \cdot \nabla w) = \int_S \xi v w \quad \forall w \in H^1_{0_{\Gamma - S}}(\Omega).$$

To approximate the problem let us use a finite element method with $u \in V_h$ continuous piecewise linear on the triangles of a triangulation of Ω:

$$\min_{v \in V_h} J_h(v) = \int_\Omega [(u - u_d)^2 + |\nabla(u - u_d)|^2] : \int_\Omega (uw + \nabla u \cdot \nabla w) \tag{9.16}$$

$$= \int_S \xi v w, \quad \forall w \in V_h,$$

where V_h is an approximation of $H^1_{0_{\Gamma - S}}(\Omega)$. Then the discrete optimality conditions are obtained in the same way:

$$\delta J = \int_S \xi(u - u_d)\delta v.$$

The Schwarz algorithm Let $\Omega = \Omega_1 \cup \Omega_2$. Let $\Gamma = \partial\Omega$ and $\Gamma_{ij} = \partial\Omega_i \cap \partial\Omega_j$. The multiplicative Schwarz algorithm for the Laplace equation starts from a guess u_1^0, u_2^0 and computes the solution of:

$$u - \Delta u = f \text{ in } \Omega, \quad u|_\Gamma = u_\Gamma$$

as the limit in n of $u_i^n, i = 1, 2$ defined by

$$u_1^{n+1} - \Delta u_1^{n+1} = f \text{ in } \Omega_1,$$

$$u^{n+1}|_{\Gamma\cap\overline{\Omega}_1-S} = u_\Gamma \quad u_1^{n+1}|_{\Gamma_{12}} = u_2^n \quad \frac{\partial u_1^{n+1}}{\partial n}|_S = \xi v$$
$$u_2^{n+1} - \Delta u_2^{n+1} = f \text{ in } \Omega_2,$$
$$u^{n+1}|_{\Gamma\cap\overline{\Omega}_2-S} = u_\Gamma \quad u^{n+1}|_{\Gamma_{21}} = u_1^n \quad \frac{\partial u_2^{n+1}}{\partial n}|_S = \xi v.$$

Derivative of the discrete problem The discretized problem is

$$\min_{v \in V_h} J_h^N(v) = \|u^N - u_d\|_\Omega^2 : \quad u_j^0 = 0, \quad n = 1..N \quad \forall w \in V_h$$
$$u_j^n|_{\partial\Omega_{ij}} = u_j^{n-1} \quad \int_{\Omega_j} [u_j^n w + \nabla u_j^n \nabla w] = \int_S \xi v w$$

where N is the number of Schwarz iterations. The exact discrete optimality conditions of the discrete problems are:

$$p^N - \Delta p^N = 2(u^N - u_d) \quad p^{N-1} - \Delta p^{N-1} = 0 \quad p_{\Gamma_{ij}}^{N-1} = p^N...$$

These are difficult to implement because we must store all intermediate functions generated by the Schwarz algorithm and integrate the system for p^n in the reverse order. So here we will try to compute approximate optimality functions; the criterion for mesh refinement will be based on the v-derivative of J

$$\theta_h = -\|u - u_d\|_S,$$

where u_h is computed by N iterations of the Schwarz algorithm. Now, using the Schwarz algorithm in the steepest descent algorithm with the Armijo rule for mesh refinement and approximate gradients, we have the following result.

9.6.2 Algorithm

```
while h > h_min
{
while |grad_{zN} J^m| > εh^γ
{
try to find a step size ρ with
```

$$-\beta\rho|\text{grad}_{zN} J^m|^2 < J(z^m - \rho\text{grad}_{zN} J^m) - J(z^m)$$
$$< -\alpha\rho|\text{grad}_{zN} J^m|^2$$

```
    if success then
{z^{m+1} = z^m - ρgrad_{zN} J^m;  m := m + 1}
else   N := N + K;
}
h := h/2;  N := N(h);
}
```

Here z is the boundary control v, J is the cost function defined by (9.16) and the gradient $\text{grad}_{zN} J$ is $u - u_d$ computed by N iterations of the Schwarz algorithm.

Let us denote by a, b the boundary of the domain Ω_1, c, d the boundary of Ω_2; a is inside Ω_2 and c is inside Ω_1, and let S be the boundary of a small circle. Functions named with capital letters are defined on Ω_1, others on Ω_2.

The following gives a complete description of the algorithm used. The test for increasing the number of the Schwarz algorithm [6] is based on the decrease of cost function being less than $0.1(0.8)^N$ and the mesh refinement, controlled by an integer giving $h = O(1/n)$, is based

(i) either on the norm of the approximate gradient of J_h, the function θ_h being less than $\epsilon(n) = 10^{-n}$,

(ii) or on the decrease of the cost function being greater than $0.001(0.8)^n$.

Naturally other choices are possible.

We give below the program in the freefem++ language [1].

```
U=10; u=1; P=10; p=0;
for( int k=0; k<=N; k++){ AA; A; BB; B; j=1;}
uu = u*om+U*OM-(u+U)*C/2;
pp = p*om+P*OM-(p+P)*C/2;
plot(sh,uu,value=1);
crit = int2d(sh)((uu-ue)^2 + (dx(uu)-dxue)^2 + (dy(uu)-dyue)^2);
real LL= 1e-10+int2d(sh)(pp^2+dx(pp)^2+dy(pp)^2),
K= int2d(sh)((uu-ue)*pp+(dx(uu)-dxue)*dx(pp)
   +(dy(uu)-dyue)*dy(pp)),
rho=abs(K/LL);
if(crit > critold-0.1/sqrt(real(N)))
{N=N+1; cout<<"N="<<N<<endl; critold=1e30;}
else {
v = v - rho*(uu-ue);
gradJ=int1d(sh,f)((uu-ue)^2);
critold=crit; // -rho*gradJ/4 -K*K/LL/4;
}
} else{
gradJ=eps; n=n+1;
th = buildmesh( a(5*n) + a1(5*n)
   + b(5*n) + c(10*n) + f(-15*n));
TH = buildmesh( e(5*n) + e1(25*n));
sh=buildmesh(e(5*n)+a(5*n)+a1(5*n)
   +b(5*n)+c(10*n)+e1(25*n)+f(-15*n));
uu=uu; v=v; w = uu-ue;
eps = eps/10; j=0; critold=1e20;
;}
criter<<-log(crit)<<endl;
 gradient<<-log(gradJ)<<endl;
schwarz<<N+1<<endl;
 nodes<<n<<endl;
```

```
epsilon<<-log(eps)<<endl;
;}
w=u-ue; plot(w,value=1);
```

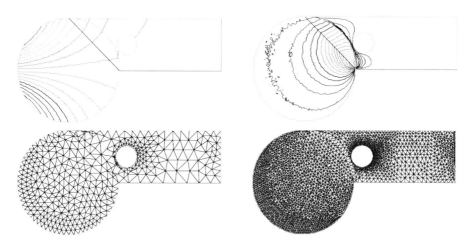

FIG. 9.8. These plots show the computed solution u (top/left) and the error $(u - u_d)$ (top/right) (maximum relative error 0.01); the second (bottom/left) and seventh (bottom/right) meshes.

In this algorithm the optimal step size is computed exactly because the criterion is quadratic. Note that the integrals are computed on the whole domain so uu, pp denotes either u, p or U, P or $(u+U)/2, (p+P)/2$ depending on whether we are in Ω_1, Ω_2 or $\Omega_1 \cap \Omega_2$. Figure 9.8 shows the computed solution u, the error $u - u_d$ and the second and seventh finite element meshes (the last is the ninth). Both meshes are generated automatically by a Delaunay-Voronoi mesh generator from a uniform distribution of points on the boundaries.

9.6.3 Numerical results

The domain is made of the unit circle plus a quadrangle made by the rectangle $(0,3) \times (0,1)$ minus the unit triangle and minus a disk S. The function which is to be recovered by the optimization process is $u_d = e^{-x\sqrt{2}} \sin(y)$. The weight on the control has been purposely chosen with oscillations: $\xi = \sin(30 * (x - 1.15)) + \sin(30 * (y - 0.5))$. We have an automatic mesh generator controlled by a parameter n and the number of points on the boundary is proportional to n. The number of Schwarz iterations is initialized at 1.

The results are shown on Fig. 9.9. After 30 iterations the gradient is 10^{-6} times its initial value, while without mesh refinement it has been divided by 100 only (multigrid effect).

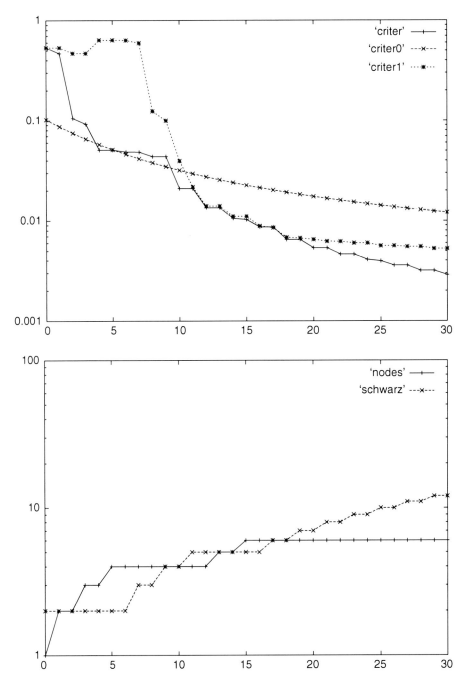

FIG. 9.9. Upper: evolution of the cost function J without mesh refinement (curve criter0) and with mesh refinement either on the norm of the gradient (curve criter1) or on the decrease of the cost function (curve criter). Lower: number of Schwarz iterations N and mesh nodes n versus iteration numbers.

9.7 Conclusion

It is a well known rule in optimization to use the exact gradient of the approximate problem rather than the approximate gradient of the exact problem. We have shown here that mesh refinement within the optimization loop enables us to slacken this rule. Our motivation is two-fold; first, there are problems where the exact gradient of the approximate problem cannot be computed simply; secondly, there is a multi-grid effect in combining mesh refinement with the descent algorithm which results in an order of magnitude speed-up.

9.8 Hypotheses in Theorem 9.3.2.1

Before closing this chapter, we would like to verify the hypothesis of Theorem 9.3.2.1.

9.8.1 *Inclusion*

If $L_{h'}$ is obtained by $q^{i_1}, ..., q^{i_{n'_f}}$ and this set contains the vertices that yielded L_h then obviously

$$L_h \subset L_{h'} \subset H_0^2(I).$$

9.8.2 *Continuity*

From [5], we know about the continuous dependence of the solution of a PDE with respect to data:

$$\alpha^n \xrightarrow{H_0^2(I)} \alpha, \ \Rightarrow u^n \xrightarrow{H^1(\Omega)} u, \ \Rightarrow J^n \to J.$$

Similarly in the discrete case, the spline is continuous with respect to the vertex position so

$$q^{i\ n} \to q^i \Rightarrow \alpha_h^n \xrightarrow{H_0^2(I)} \alpha_h, \ \Rightarrow u_h^n \xrightarrow{H^1(\Omega)} u_h, \ \Rightarrow J_h^n \to J_h.$$

9.8.3 *Consistency*

It is obvious that

$$\forall \alpha, \ \exists \alpha_h \to \alpha \ \text{with} \ J_h \to J,$$

because one just puts vertices on the continuous boundary, applies the construction, and, as the number of vertices increases, the discrete curve converges to the continuous one if the following is observed:

- Corners of the continuous curve are vertices of the discrete curves.
- The distance between boundary vertices converges uniformly to zero.

9.8.4 *Continuity of θ*

Conjecture There exists ε such that $\alpha \in H_0^2 \ \Rightarrow u \in H^{3/2+\epsilon}(\Omega)$.
  Arguments We know that $\alpha \in C^{0,1} \ \Rightarrow u \in H^{3/2}$ and $\alpha \in C^{1,1} \ \Rightarrow u \in H^2$ [4]. This technical point of functional analysis is needed for the continuity of θ. If so the following convergence properties hold

$$\alpha^n \xrightarrow{H^2} \alpha, \quad \Rightarrow u^n \xrightarrow{H^{3/2+\epsilon}(\Omega)} u, \quad \Rightarrow u^n|_\Sigma \xrightarrow{H^{1+\epsilon}} u|_\Sigma,$$

$$p^n \xrightarrow{H^{3/2+\epsilon}(\Omega)} p, \quad \Rightarrow p^n|_\Sigma \xrightarrow{H^{1+\epsilon}} p|_\Sigma,$$

$$\Rightarrow \nabla u^n \nabla p^n|_\Sigma \xrightarrow{L^2(\Sigma)} \nabla u \nabla p|_\Sigma.$$

9.8.5 Continuity of $\theta_h(\alpha_h)$

Recall that a variation $\delta\alpha_h$ (i.e. a boundary vertex variation $\delta q^j, j \in \Sigma$) implies variations of all inner vertices $\delta\alpha, \delta q^k, \forall k$. The problem is that θ is a boundary integral on Σ and θ_h is a volume integral! We must explain why:

$$\delta J_h - \int_\Omega (\nabla u_h (\nabla \delta q_h + \nabla \delta q_h^T) \nabla p_h - \nabla u_h \cdot \nabla p_h \nabla \cdot q_h)$$
$$+ 2\epsilon \int_\Sigma \frac{d^2\alpha}{ds^2} \frac{d^2\delta\alpha}{ds^2} \xrightarrow{?} \delta J = - \int_\Sigma \delta\alpha \left(up + \nabla u \nabla p + 2\epsilon \frac{d^4\alpha}{ds^4} \right).$$

This is the object of Lemma 9.5.8.1.

9.8.6 Convergence

This comes from the theory of finite element error analysis [2]:

9.8.6.1 Lemma

$$\left| \int_\Sigma \nabla u_h \nabla p_h - \nabla u \nabla p \right| \leq Ch^{1/2}(\|p\|_2 + \|u\|_2)$$

with the following triangular inequalities

- $|a_h b_h - ab| = (a_h - a)(b_h - b) + b(a_h - a) + (b_h - b)a$
  $\leq |b||a_h - a| + |a||b_h - b| + |a_h - a|^2 + |b_h - b|^2$
- $|\nabla u_h - \nabla u|_{0,\Sigma} \leq |\nabla(u_h - \Pi_h u)|_{0,\Sigma} + |\nabla(\Pi_h u - u)|_{0,\Sigma}$

plus an inverse inequality for the first term and an interpolation for the second.

References

[1] Bernardi, D. Hecht, F. Otsuka, K. Le Hyaric, A. and Pironneau, O. (2006). *Freefem++, userguide*, www.freefem.org.
[2] Ciarlet, Ph. (1978). *The Finite Element Method for Elliptic Problems*, North-Holland, New York.
[3] Dicesaré, N. Pironneau, O. and Polak, E. (1998). Consistent approximations and optimal shape design, *report LAN-UP6*, **98001**.
[4] Grisvard, P. (1985). *Elliptic Problems in Non-Smooth Domains*. Pitman, New York.
[5] Ladyzhenskaya, O. and Ural'tseva, N. (1968). *Linear and quasilinear elliptic equations*, Academic Press, New York.

[6] Lions, P.-L.. (1998). On the Schwarz alternating method. I,II,III. Lectures on Domain decomposition Methods, *Proc. Int. Symp. on DDM*, Springer-Verlag, Berlin.

[7] Pironneau, O. (1984). *Optimal Shape Design of Elliptic Systems,* Springer-Verlag, New York.

[8] Pironneau, O. and Polak, E. (2002). Approximate gradient for control problems, *SIAM J. Optim. Control*, **41(2)**, 487-510.

[9] Polak, E. (1997). *Optimization: Algorithms and Consistent Approximations,* Springer, New York.

10
NUMERICAL RESULTS ON SHAPE OPTIMIZATION

10.1 Introduction

In this chapter we present some constrained shape optimization problems at various Mach numbers ranging from incompressible flows to supersonic. All these computations have been performed on standard personal computers. The overall calculation effort even for 3D configurations is a matter of hours. This includes mesh adaptations, gradient computations and flow solutions. When mesh adaption is used during optimization a final computation is performed on the final shape. This is to make sure that the solutions is fully converged and mesh-independent.

We have chosen not to report all the examples presented in the first edition of the book, but rather retain the most insightful configurations. In particular, we want to illustrate the application of incomplete sensitivities in some important classes of problems such as drag reduction under constraints on the lift and the volume. The chapter also aims at showing how to reformulate the problem in cases not suitable for direct application of incomplete sensitivities. This reformulation often tells us a lot about the initial problem itself. Several situations where incomplete sensitivity cannot be originaly applied are considered: shape design in radial and axial turbomachines and sonic boom reduction. The reformulation aims at using either the governing equations or extra engineering information to link all involved quantities in the functional and constraints to shape-based quantities such as aerodynamic coefficients or purely geometric quantities.

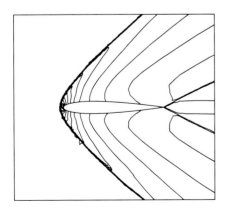 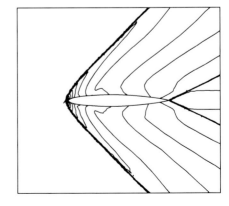

FIG. 10.1. Supersonic drag reduction. Iso-Mach evolution (initial and step 11).

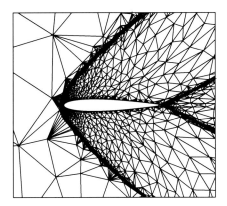 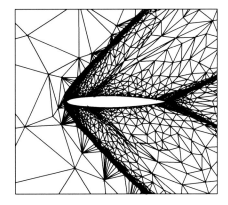

Fig. 10.2. Supersonic drag reduction. Mesh evolution (initial and step 11).

10.2 External flows around airfoils

We consider a drag reduction problem under volume and lift constraints. The flow Mach number is 2. The initial profile is a NACA0012 (Figs. 10.1 and 10.2). For the shock to be attached, the leading edge has to be sharp. But the initial shape has a smooth leading edge. This means that the optimization procedure has to be able to treat the apparition of singular points. This cannot be done if we use a CAD-based parameter space (splines for instance).

The drag has been reduced by about 20% (from 0.09 to 0.072) as the shock is now attached while the volume has been conserved (from 0.087 to 0.086). The initial lift coefficient should be zero, as the airfoil is symmetric. Due to numerical errors, the lift varies from -0.001 to 0.0008. The final shape is almost symmetric without specific prescription.

Results for a similar constrained drag reduction problem with mesh adaptation have been shown in Section 9.5.13 for a RAE2822 profile in the transonic regime at Mach number 0.734 and Reynolds number 6.510^6 and at an incidence of $2.79°$ (see Fig. 10.3) [8, 7].

10.3 Four-element airfoil optimization

Let us now consider a multi-component shape. The initial airfoil is a four-element profile at inflow Mach number 0.2 and $4°$ of incidence (see Fig. 10.4 and 10.5). The aim is to increase the lift at given drag and volume. One can also consider a maximization of the lift-to-drag ratio at given volume.

The ratio has almost doubled after optimization and the volume conserved. This result is, however, unphysical as the drag here is only due to numerical dissipation. Indeed, for this subsonic inviscid configuration we should have a zero drag solution. This is a classical problem with upwind solvers especially when the mesh is coarse. We saw in chapter 4 that for these high incidence

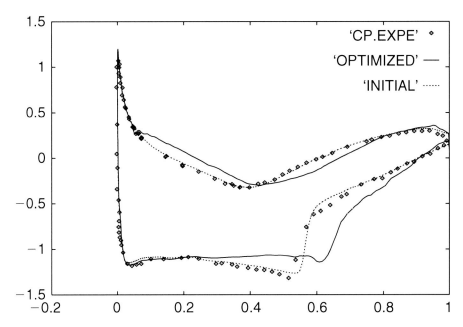

FIG. 10.3. Transonic turbulent flow. Pressure coefficients over the initial and optimized shapes vs. experimental data.

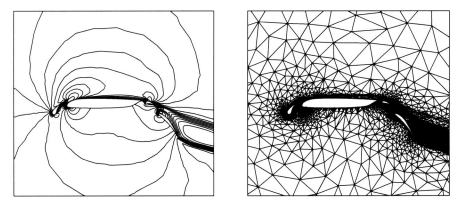

FIG. 10.4. Four-element airfoil optimization. Final iso-Mach contours and adapted mesh.

configurations we need turbulence models and accurate time integration as the flow is unsteady.

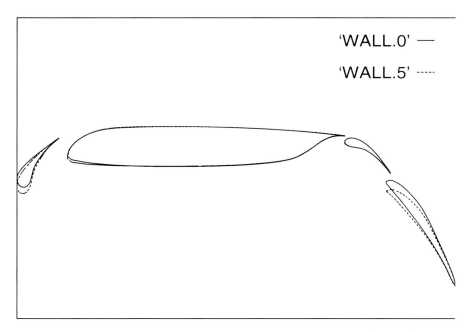

FIG. 10.5. Four-element airfoil optimization. Initial and final (step 5) shapes.

10.4 Sonic boom reduction

Sonic boom and fuel consumption reduction are important research issues for the qualification of supersonic civil transport aircraft. Fuel consumption reduction can be seen as drag reduction at a given volume and lift, for instance. Sonic boom concerns both noise generation over land and its impact on marine life. Knowing the pressure distribution at a given distance below the aircraft, we can find the pressure signature on the ground using nonlinear wave propagation methods [13, 11].

The characteristics of the problem are mostly inviscid [2, 12, 4]. The sonic boom paradox makes the optimization problem difficult as drag and sonic boom reductions are incompatible. Indeed, at high speed, drag reduction leads to sharp leading edges and attached shocks, while to reduce the sonic boom the shape needs to be smooth and the shock to remain bow (see Fig. 10.6).

This problem is also interesting as the sonic boom is defined on a boundary away from the wall. Therefore, the application of the incomplete gradient formulation is not straightforward and needs a redefinition of the problem.

To illustrate this problem we consider a 2D model problem for sonic boom where the aim is to reduce the pressure jump on the lower boundary of the computational domain and also the drag under volume constraint. The geometry is a 2D fore-body at zero incidence and the inflow Mach number is 2.

We have noticed that to reduce the pressure jump along this boundary it is sufficient to keep the leading edge smooth while doing drag reduction (see Fig.

10.7) introducing the following requirement:
- specify that the wall has to remain smooth;
- ask for the local drag force around the leading edge to increase while the global drag force decreases.

In other words, incomplete sensitivity is not evaluated for the following functional:
$$J(x) = C_d + |\text{Vol} - \text{Vol}^0| + |\Delta p|,$$
where Δp is the pressure rise along the lower boundary of the computation domain. Instead, incomplete sensitivity calculation is made for:
$$J(x) = C_d + |\text{Vol} - \text{Vol}^0| + \frac{1}{C_d^{loc}},$$
where C_d^{loc} is the drag force generated around the leading edge.

Minimization using descent directions based on incomplete sensitivity for this new definition of the functional shows that C_d^{loc} is increased by about 10% the leading edge remaining smooth while the global drag decreases (see Fig. 10.7). This also corresponds to a reduction of the pressure rise on the lower boundary.

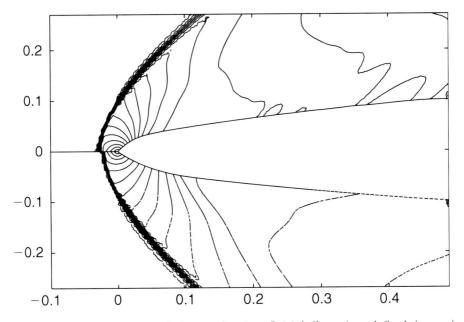

FIG. 10.6. Sonic boom and drag reduction. Initial (lower) and final (upper) iso-Mach contours. The shock remains bow and is even pushed away from the wall.

This idea has been extended to a full aircraft [6] (see Fig. 10.8). We consider a supersonic business jet geometry provided by the Dassault Aviation company

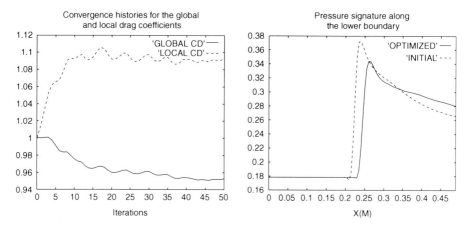

FIG. 10.7. Sonic boom and drag reduction. Global and local drag coefficients evolution and the pressure rise along the lower boundary.

(see Fig. 10.9). The cruise speed is Mach 1.8 at zero incidence and the flight altitude is 55000 ft.

One difficulty in sonic boom prediction for 3D geometries is the definition of the pressure signature on the ground. This is why usually only a 1D signature (along the trace of the centerline of the aircraft on the ground) is analyzed. This is because in cruise conditions maximum boom is observed along the centerline. In fact, even this 1D trace on the ground is difficult to obtain because of numerical diffusion due to having not enough refined meshes away from the aircraft. Mesh adaptation is a powerful tool to remove this difficulty [1]. For extension to 3D geometries we need to define C_d^{loc} evaluated over the leading edge regions Γ_{loc} which is defined as where $u_\infty . \vec{n} < 0$, u_∞ being the inflow velocity and $\vec{n}$ the local normal to the shape.

Figure 10.10 shows a cross-section of the close-field CFD pressure signature in the symmetry plane and after propagation with the wave model. Figure 10.11 shows iso-contours of normal deformations with respect to the original shape. The near-field initial pressure rise (IPR) is higher for the optimal shape but lower after propagation: it is not enough to control the IPR to reduce the boom. In this optimization the drag is reduced by 20% while the lift is increased by 10% and the geometric constraints based on local thickness and overall volume conservations are satisfied.

10.5 Turbomachines

We show how the incomplete sensitivity approach can be adapted to rotating blades, both axial and radial. Axial blades are mainly used for air and radial blades for heavier liquids such as water, wine and oil. Because of this difference in weight for the liquids (basically more that 1000 times) a one percent gain in efficiency means a lot in terms of energy consumption for radial blades. We

218   *Numerical results on shape optimization*

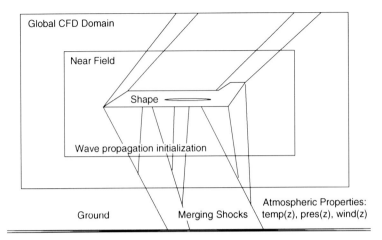

FIG. 10.8. Shock wave pattern and illustration of the near-field CFD computation domain and the initialization of the wave propagation method with CFD predictions. The coupling should be done far enough from the aircraft, justifying the mentioned need for mesh adaptation.

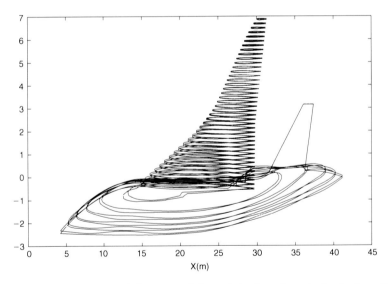

FIG. 10.9. By-section definition of the original and optimized aircrafts.

especially discuss the case of blade cascades but what is said here can be applied to any axial or radial stator or rotor.

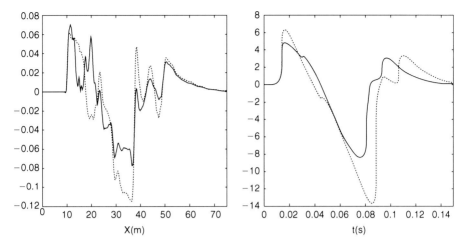

FIG. 10.10. Cross-section of the near-field CFD pressure variations in the symmetry plane (left) and after propagation (right) for the initial (dashed curves) and optimized (continuous curves) shapes.

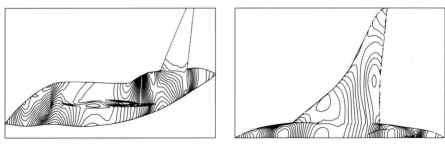

FIG. 10.11. Iso-contours of normal deformation with respect to the original shape.

10.5.1 Axial blades

Let us consider first the design of axial blades [5, 9]. Such blades are used in ventilation systems. A ventilation system is defined by its number of blades N and by its running characteristics. The blade cascade is obtained by considering cross-sections of the ventilation system for a given radius $R_{\min} < R < R_{\max}$. Periodicity allows us to consider only one blade (see Fig. 10.12 and 10.13).

The other characteristics are the air debit q, the rotating velocity ω, the profile chord c, the distance between two blades $t = 2\pi R/N$, the angle λ between the profile chord and the x axis, the angles β_∞, β_1 and β_2 between, respectively, the inflow velocity and the normal to the leading and trailing edges and the x axis, the incidence angle $\alpha = \beta_\infty - \lambda$ and the deflection angle $\theta = \beta_2 - \beta_1$. The inflow and outflow boundaries are chosen such that the inflow remains constant and the

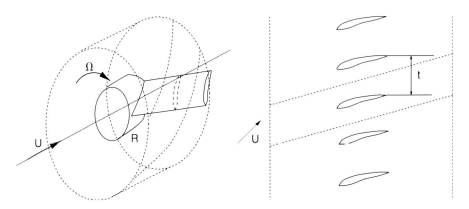

FIG. 10.12. Blade cascade obtained at radius R.

outflow velocity has the same angle along it. In other words, these boundaries have to be far enough from the blade.

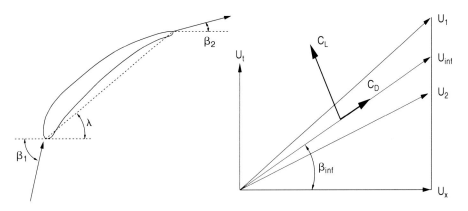

FIG. 10.13. A typical blade and how to define the aerodynamic forces.

The inflow conditions are given by the air debit and the rotation velocity (see Fig. 10.15):

$$\begin{cases} U_x = \frac{q}{\pi(R_{max}^2 - R_{min}^2)}, \\ U_t = \omega R. \end{cases} \quad (10.1)$$

The aerodynamic forces C_l, C_d, and C_m are expressed in the frame attached to the inflow velocity (see Fig. 10.14). The efficiency of a ventilation system is defined by:

$$\eta = \frac{q\triangle p}{\omega C_m}, \quad (10.2)$$

where Δp is the pressure difference between the inlet and outlet boundaries. Optimizing a ventilation system therefore means increasing its efficiency. This means for a given debit, keeping constant or increasing the pressure difference between inlet and outlet and reducing the aerodynamic moment. In a car cooling device, for instance, this implies a reduction in fuel consumption as well as noise.

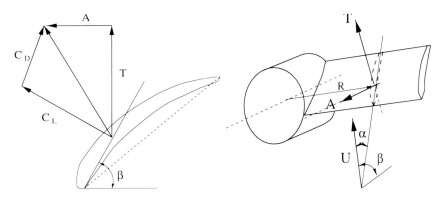

FIG. 10.14. Aerodynamic efforts on the blade.

In this case we therefore have a special constraint due to the fact that we want to prescribe a given difference of pressure between the inlet and outlet. This is a classical constraint for internal flow applications. Again, this is not suitable for the incomplete sensitivity method as one needs a cost function and constraints to be described through boundary integrals over the shape, while this constraint involves inflow and outflow boundaries. In order to use our approximation, we need therefore to express this constraint through boundary integrals. Integrating by parts the steady momentum equation, we have:

$$\int_\Gamma u(u.n)d\sigma + \int_\Gamma Tn d\sigma = 0,$$

where $T = -pI + (\nu + \nu_t)(\nabla u + \nabla u^T)$ is the Newtonian stress tensor.

A first approximation is to neglect viscous effects at the inlet and outlet boundaries (see Fig. 10.15). Then, using the periodicity conditions for lateral boundaries, we have:

$$\int_{\Gamma_i} u(u.n)d\sigma + \int_{\Gamma_o} u(u.n)d\sigma + \int_{\Gamma_i} pn d\sigma + \int_{\Gamma_o} pn d\sigma + \int_{\Gamma_w} Tn_w d\sigma = 0. \quad (10.3)$$

Let us denote mean value quantities at the inlet and outlet boundaries by u_i, u_o, p_i, and p_o. The inlet and outlet boundaries have the same length L. From

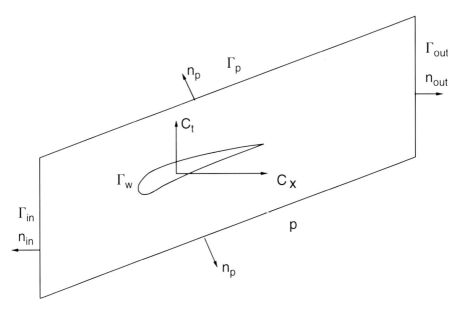

FIG. 10.15. Blade cascade computational domain.

the continuity equation ($\nabla.u = 0$) and due to periodicity on lateral boundaries and the slip condition on the shape, we have $u_i.n_i = -u_o.n_i$.

The first component (along the x axis) of (10.3) is therefore reduced to:

$$\triangle p = p_o - p_i = -C_x \frac{1}{2}\rho_\infty |u_\infty|^2 \frac{c}{L} \qquad (10.3)$$

where $C_x = C_d \cos(\beta) - C_l \sin(\beta)$ is the horizontal aerodynamic force. Hence, the pressure difference between inlet and outlet boundaries can be expressed through the horizontal aerodynamic force on the blade which is a boundary integral making possible the application of incomplete sensitivities. This approach has been numerically and experimentally validated [9] for blade cascades. With incomplete sensitivities the cost of the optimization is about 1.5 times the cost of the initial flow evaluation. It is observed that the reformulation of the functional is effective: the cost function is decreasing while the pressure difference between inlet and outlet is increasing.

10.5.2 Radial blades

The previous analysis can be extended to radial blades used in centrifugal pumps. Indeed, in radial turbomachinery, the pressure difference between inlet and outlet should be conserved or increased, while the torque should be reduced by the design [3]. Unlike axial blades here the rotation vector is orthogonal to the blade axis and flow moves away from the rotation axis. In radial turbomachines the main shape parameter is the spanwise blade angle distribution from leading to

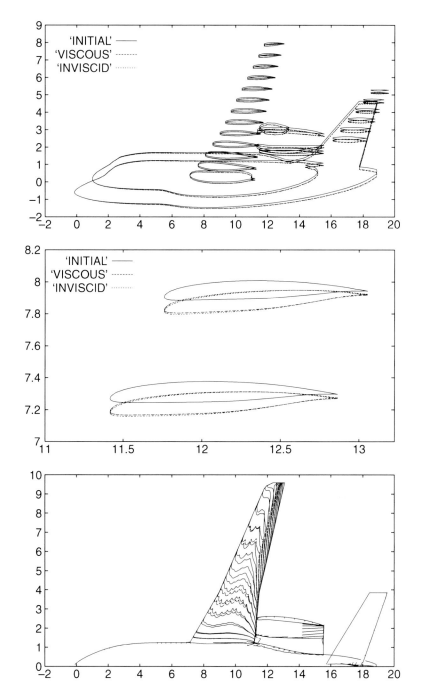

FIG. 10.16. Viscous vs. inviscid states for 3D optimization. By-section definitions of the shape used for maximum thickness constraint (top and middle). Iso-contours of the normal deformation to the shape for the two optimizations (lower).

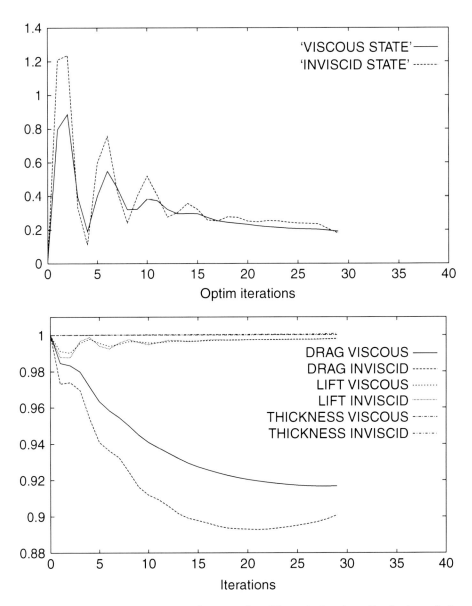

FIG. 10.17. Viscous vs. inviscid states for 3D optimization. Evolution of the incidence correction for lift recovery (upper). Convergence histories for the normalized drag and lift coefficients and the maximum thickness constraint (lower).

trailing edges. It has been observed that blade efficiency is not very sensitive to small variations of its thicknesses [10]. This means that the volume of the blade will remain quite unchanged during design.

As for axial fans, centrifugal pumps are designed for a given design point and a hydraulic efficiency is defined by (10.2). We therefore need to express the pressure difference between inlet and outlet in terms of boundary integrals over the shape involving products of state by geometry quantities. The flow rate is enforced through the inlet boundary condition on the velocity.

Neglecting viscous effects and working with the momentum equation for rotating flows one can show that, again, maintaining unchanged the radial and axial forces on the blade will imply the pressure difference between the inlet and outlet to remain unchanged [3].

10.6 Business jet: impact of state evaluations

We saw in Chapter 2 that transonic viscous flow over airfoils has a shock appearing earlier on the profile than the same flow computed without viscous effects included, and this impacts the final shape after optimization. We would like to see the impact of state evaluation on the optimization for a full aircraft in the transonic regime. Viscous simulations use a $k - \varepsilon$ turbulence model with wall functions. We consider the same drag reduction problem as in (10.2) under lift, volume, and also thickness constraints.

The inflow Mach number is 0.8 with an incidence of 1.5 degrees. The Reynolds number is 510^6 for viscous simulations and corresponds to a cruise altitude of 40 000 feet. We perform two optimizations with incomplete gradients based on inviscid and viscous state evaluations. The optimization concerns only the wing. There are about 10 000 optimization points over the wing. Both the leading and trailing edges are let free to move and no limitation is introduced for the deformations. The approach confirms the expected wing spanwise twist without any dedicated parameterization (see Fig. 10.16). The drag has been reduced by about 10% while the lift, the volume and the local by-section maximum thickness are almost unchanged for both optimizations. The final shapes are slightly different both chordwise and spanwise (see Fig. 10.16). The deformation is clearly three dimensional. Lift is recovered by incidence correction and the thickness as mentioned in Chapter 6 (see Fig. 10.17).

References

[1] Alauzet, F. (2006). Adaptive sonic boom sensitivity analysis, *Proc. Eccomas-CFD06*, Springer, Berlin.

[2] Alonso, J. Kroo, I. and Jameson, A. (2002). Advanced algoritms for design and optimization of QSP, *AIAA*, **02-0144**.

[3] Derakhshan, S. Nourbakhsh, A. and Mohammadi, B. (2008). Incomplete sensitivity for 3D radial turbomachine blade optimization, *Comput. Fluids*, **37(10)**, 1354-1363.

[4] Farhat, C. Maute, K. Argrow, B. and Nikbay, M. (2002). A shape optimization methodology for reducing the sonic boom initial pressure rise, *AIAA*, **02-145**.

[5] Medic, G. Mohammadi, B. Moreau, S. and Stanciu, M. (1998). Optimal airfoil and blade design in compressible and incompressible flows, *American Institute of Aeronautics and Astronautics*, **98-124**.

[6] Mohammadi, B. (2004). Optimization of aerodynamic and acoustic performances of supersonic civil transports, *Internat. J. Num. Meth. Heat Fluid Flow*, **14(7)**, 893-909.

[7] Mohammadi, B. (1997).Practical applications to fluid flows of automatic differentiation for design problems, *VKI lecture series*, **1997-05**.

[8] Mohammadi, B. (1996). Optimal shape design, reverse mode of automatic differentiation and turbulence, *American Institute of Aeronautics and Astronautics*, **97-0099**.

[9] Stanciu, M. Mohammadi, B. and Moreau, S. (2003). Low complexity models to improve incomplete sensitivities for shape optimization, *IJCFD*, **17(1)**, 1-13.

[10] Stepanoff, A.J. (1957). *Centrifugal and Axial Flow Pumps*, John Wiley, New York.

[11] Thomas, Ch. (1972). Extrapolation of sonic boom pressure signatures by the waveform parameter method, *NASA TN*, **D-6832**.

[12] Vazquez, M. Koobus, B. and Dervieux, A. (2004). Multilevel optimization of a supersonic aircraft, *Finite Elem. Anal. Design*, **40(15)**, 2101-2124.

[13] Whitham, G. B. (1952). The flow pattern of a supersonic projectile, *Comm. Pure Appl. Math*, **5(3)**, 301-348.

11
CONTROL OF UNSTEADY FLOWS

11.1 Introduction

In this chapter we present some results for passive and active flow control problems with gradient based control laws in fixed and moving domains. We already mentioned this issue in Chapter 8 where an example of shape optimization for an unsteady flow has been discussed. In the aeroelastic cases, we have chosen the structural characteristics in order to have resonance. One difficulty in multidisciplinary problems involving fluid and structure is the non-conformity of interface discretizations used in the different models. The coupling algorithms are presented using simple flow models. The extra cost compared to the original uncontrolled simulations is less than 5%.

Control of unsteadiness in flows is of great importance for industrial applications. For instance, in fluid structure interaction problems it might lead to a resonance with the structure eigen-modes. Therefore, the prediction and control of such flows is useful.

Passive control is currently in use in aeronautics to delay transition to turbulence and reduce the drag. The procedure is simple and consists in adding a suction device in the leading edge region of wings (the amount of suction is less than 0.1% of the inflow velocity). The suction intensity is mainly constant in time and has been tuned for commercial aircraft for cruise condition. Active control is not yet at the same level of industrial development. In aeronautics it also implies security and qualification problems.

Both passive and active control problems can be reformulated as unsteady cost function minimization problems. We can define control laws from gradient methods based on exact or partial gradient evaluations. We aim to show how to use a gradient-based shape optimization tool, first designed for steady configurations, as a tool for flow control.

We saw in Chapter 7 that different control laws can be obtained depending on the minimization algorithm we use and more precisely on the corresponding dynamical system. Hence, different active or passive control laws can be derived but none of them are optimal except if the minimization algorithm can reach the global minimum of the minimization problem.

One advantage of the gradient-based control law is that it does not require any evaluation of the flow to build the transfer function. Indeed, the control law is built in real time and is self-adaptive: if we change the Reynolds or Mach numbers or even the shape, the same gradient-based law will produce a new control.

To make the control cheap, an important issue is then the validity of incomplete sensitivities in time-dependent problems. We recall that the key idea in incomplete gradient evaluation is that the sensitivity with respect to the state can be neglected in the gradient for cost functions defined on the shape and involving products of state by geometry quantities. Incomplete sensitivity makes it possible to perform analysis and control at the same time (as no adjoint problem is solved). In Chapter 8 we saw how to implement small shape deformations using equivalent injection/transpiration boundary conditions [5–7]. Indeed, for a control law to be efficient and realizable the amount of the required deformation or equivalent injection/suction velocity has to be as small as possible. For this reason, transpiration boundary conditions are more suitable for such simulations than Arbitrary Lagrangian Eulerian (ALE) formulations, for instance. ALE formulations introduce extra uncertainties due to the mesh movement based on spring type models solved with iterative solvers. These perturbations can be of the same order, or bigger, than the amount of the control introduced. Another argument in that sense comes from the fact that in control we would like to act on small eddies with high frequencies introducing small amounts of energy, but ALE rather considers flow structures with long time-scales which can be seen by the structure. In fact, a combination of ALE techniques and transpiration boundary conditions is necessary to efficiently take into account both aspects.

A similar remark can be made when using unsteady RANS models to simulate turbulent flows. Indeed, as we said, we are aiming to act on small eddies which drop, in principle, into the part of the spectrum modeled by $k - \varepsilon$ type models. For this reason, we think that LES models should be used to avoid losing this information. But in that case, the uncertainties due to boundary conditions should be better understood in LES models, especially for external flows.

11.2 A model problem for passive noise reduction

Consider the following minimization problem:

$$\min_{x \in \mathcal{O}_{ad}} J(x(y), y, t, u(x(y), y, t)). \tag{11.1}$$

The state equation is a modified Burgers equation

$$u_t + 0.5(u^2)_y - \nu u_{yy} = (\mu y + g(y,t) + x(y))u, \quad \text{on } \mathcal{D} =]-1,1[\tag{11.2}$$

$$u(t,-1) = u_l, \quad u(t,1) = u_r, \quad u(t=0,y) = u_0(y).$$

Here the control is a function $x(y)$ in space only and it does not depend on time. It can be seen as a passive control defined once for all. Our aim is to reduce state fluctuations in time. Of course, the unsteadiness can be completely removed only if $x = g(y,t)$, but then the control would depend on time. We consider the following configuration: $\mu = 0.3, \nu = 10^{-4}, g(y,t) = 0.5\sin(10\pi t)\cos(6\pi t)y$. The equation is discretized on a uniform mesh with 101 nodes with a classical explicit RK3 scheme in time and a consistent numerical viscosity is used for stabilization. The admissible control space is $\mathcal{O}_{ad} = [-1,1]^{101}$.

We consider the minimization of the following functional

$$J(x(y)) = \int_{t>0} j(t)dt, \quad j(x(y),t) = \int_{\mathcal{D}} (u_t^2 + u_{tt}^2) dy. \quad (11.3)$$

One can also consider a functional expressed in frequency space such as:

$$\hat{J}(x(y)) = \int_{\omega \leq \omega_{max}} \hat{j}(x(y),\omega) d\omega \quad (11.4)$$

using the Fourier transform of the signal $j(t)$. The aim then is to minimize the overall spectral energy (or some other norm) for instance.

Figure 11.1 shows that a control obtained by minimizing (11.3) or (11.4) can be efficient in reducing the fluctuations. Here the gradient is computed using the reverse mode of automatic differentiation described in Chapter 5. The former functional also requires the linearization of a FFT series calculation in the loop linking the parameters to the functional. We voluntarily consider that point as a black-box.

11.3 Control of aerodynamic instabilities around rigid bodies

We describe how to control unsteady flows around fixed rigid bodies. Compared to the previous example here the control depends on time. The aim is to show that incomplete sensitivities are also efficient to define control laws. The control laws are prescribed via equivalent injection velocities corresponding to the shape deformation prescribed by the minimization approach (here the steepest descent method).

We have validated the approach for various flows such as flows around a circular cylinder at various Mach numbers and flows over different airfoils [3]. We show the case of a buffeting control by injection/suction.

We consider a RA-16 profile at 3.5° of incidence. The inflow Mach number is 0.723 and the chord-based Reynolds number is 4.2×10^6. The unsteady wake creates a periodic movement of the upper-surface shock (Fig. 11.2). The main frequency is at about 15 Hz. The correct prediction of this flow is a difficult task in itself. We want to reduce the shock displacement by mass injection/suction for two different locations of the injection device between $0.4 < (x/\text{chord}) < 0.6$ or $0.35 < (x/\text{chord}) < 0.65$.

The maximum value for the injection intensity corresponds to 5% of the inflow velocity. The unsteady cost function here is the lift coefficient. We aim to reduce its fluctuation. We can see that in both cases the control is efficient, but that in the first case we obtain a higher value for the resulting lift coefficient after control and the same level for the drag (see Figs. 11.3 and 11.4).

11.4 Control in multi-disciplinary context

Some problems involve several physical models. Prediction of flutter and design of the structure leading to a higher flutter speed is a good example where accurate prediction of the physics of the problem involving fluid and structure

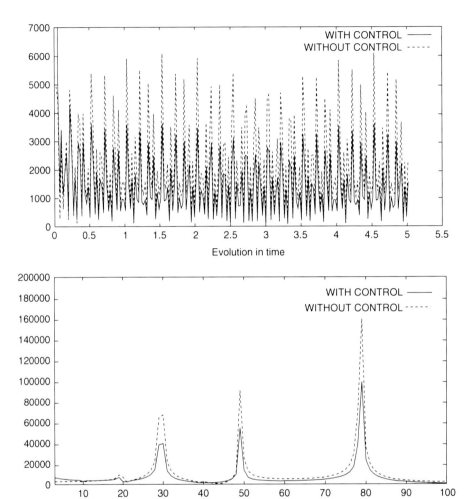

FIG. 11.1. Model problem for noise reduction. Evolution of $j(t)$ (upper) and its spectral representation on a given time window for the uncontrolled and controlled simulations of (11.2) minimizing (11.3).

characteristics is necessary for a correct design. One shows coupling strategies between two (or more) models for the flow and structure behavior together with a dynamic system used for minimization.

11.4.1 A model problem

Let us consider a one-dimensional coupling between a first-order parabolic and a second-order hyperbolic equation (e.g. an incompressible fluid and a wave-based structural model). The calculation domain is split into $\Omega_1 =]-1, S[$ for the

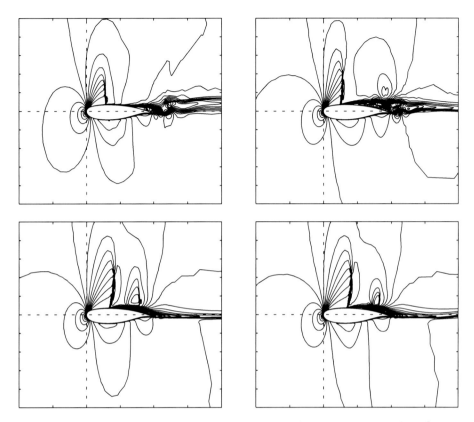

FIG. 11.2. Buffeting control. Four Mach fields for the non-perturbed flow.

parabolic equation and $\Omega_2 =\,]S,1[$ for the hyperbolic one (S is seen as the shape with $|S| \ll 1$).

The flow model (parabolic) is represented by an advection-diffusion equation for a variable u (a velocity) with $c(y) \in \mathbb{R}$ and $\mu(y) > 0$:

$$u_t + (cu)_y - (\mu u_y)_y = f(y,t), \quad \text{on } \Omega_1(t). \tag{11.5}$$

The structural model (hyperbolic) is a wave equation for U (a displacement):

$$U_{tt} - U_{yy} = u(S(t))\delta(S(t)), \quad \text{on } \Omega_2(t). \tag{11.6}$$

To be able to handle mesh deformation in the fluid domain, one also introduces the following equation for the motion of the domain Ω_1:

$$-Y_{yy} = 0, \quad \text{on } \Omega_1(t), \tag{11.7}$$

where Y is the displacement of a material point in $\Omega_1(t)$. The system above is considered with the following initial and boundary conditions:

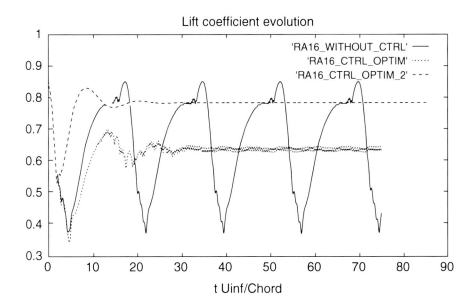

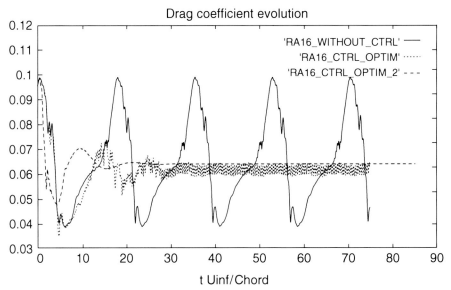

FIG. 11.3. Buffeting control. Lift and drag coefficient histories without and with our gradient-based control law for two different positions for the control devices $0.4 < x < 0.6$ and $0.35 < x < 0.65$ and with a maximum injection velocity corresponding to 5% of the inflow velocity. In this former case, not only is the unsteadiness removed but the average lift is also improved.

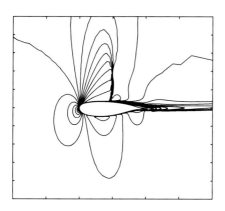 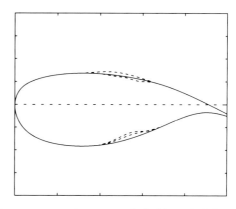

FIG. 11.4. Buffeting control. Iso-Mach contours after control (left) and different instantaneous deformations simulated by injection boundary conditions (right).

$$u(-1,t) = u_y(S,t) = 0,$$
$$U_y(S,t) = U(1,t) = 0,$$
$$Y(-1,t) = 0, Y(S,t) = U(S,t),$$
$$u(y,t) = u_0(y) \quad \text{on } \Omega_1(t),$$
$$U(y,t) = U_0(y), U_t(y,0) = U_1(y) \quad \text{on } \Omega_2(t).$$

11.4.1.1 *Handling domain deformation* To account for time-dependent domain deformations one uses the following integration rule:

$$\left(\int_{\omega(t)} g(y,t) dy \right)' = \int_{\omega(t)} g_t(y,t) dy + \int_{\partial \omega(t)} g\, \dot{y}.n\, d\sigma, \qquad (11.8)$$

where $\dot{y}$ is the speed at which the boundary moves. Integrating (11.5) on Ω_1 and using (11.8) and the divergence formula one has:

$$\left(\int_{\omega(t)} u dy \right)' + \int_{\omega(t)} (u(c-\dot{y}))_y dy - \int_{\omega(t)} (\mu u_y)_y dy \qquad (11.9)$$

$$= \int_{\omega(t)} f dy \quad \forall \omega(t) \subset \Omega_1(t).$$

ALE formulations (Arbitrary Lagrangian Eulerian) are discrete forms of this expression [7].

Discrete (11.5) One observes that domain deformations add a term into the advection operator. The speed of deformation needs to be accounted for in the upwinding procedure. In one dimensional discrete $\Omega_{1h} = \bigcup_i]y_i, y_{i+1}[$ this is easy to define at each node i:

$$\dot{y}_i = \frac{y_i^{n+1} - y_i^n}{\Delta t}.$$

Discrete (11.9) is then straightforward using finite volumes or elements if one uses the following approximation:

$$\left(\int_{\omega(t)} u\, dy\right)' \sim \frac{|\omega^{n+1}|u^{n+1} - |\omega^n|u^n}{\Delta t},$$

where $|\omega^n|$ is the length (surface in 2D, volume in 3D) of the control volume at iteration n after mass lumping.

Discrete (11.6) In the same way, the discrete wave equation in the moving domain $\Omega_{2h}(t)$ using P^1 finite elements reads:

$$\int_{\Omega_2(t)} U_{tt} w_i\, dy - \int_{\Omega_2(t)} U_{yy} w_i = \int_{\Omega_2(t)} u(S(t))\delta(S(t)) w_i\, dy. \quad (11.10)$$

Expressing U on the basis w_i, $i = 1, ..., N$ with N the number of points in $\Omega_{2h}(t)$,

$$U(y, t) = \sum_{i=1}^{N} U_i(t)\, w_i(y),$$

one has:

$$M_{i,i-1} U''_{i-1}(t) + M_{i,i} U''_i(t) + M_{i,i+1} U''_{i+1}(t) + K_{i,i-1} U_{i-1}(t) + K_{i,i} U_i(t)$$
$$+ K_{i,i+1} U_{i+1}(t) = u(S(t)) \int_{\Omega_{2h}(t)} \delta(S(t)) w_i\, dy = u(S(t)) w_i(S(t)), \quad (11.11)$$

where M and K are, respectively, the mass and rigidity matrices:

$$M_{i,j} = \int_{\Omega_{2h}(t)} w_i w_j\, dy,$$

$$K_{i,j} = \int_{\Omega_{2h}(t)} w'_i w'_j\, dy.$$

Finally, one needs a temporal discrete form for $U''(t)$. One popular choice is by an implicit second-order Newmark scheme (see below for details of coupling in time):

$$U_i^{n+1} = U_i^n + \Delta t \frac{U'^{n+1}_i + U'^n_i}{2} \quad (11.12)$$

$$U'^{n+1}_i = U'^n_i + \Delta t \frac{U''^{n+1}_i + U''^n_i}{2}.$$

The deformation $\Omega_{2h}(t^n) \to \Omega_{2h}(t^{n+1})$ is given by $y_i^{n+1} = y_i^n + U_i^{n+1}$. The coupling by the right-hand-side uses $u^{n+1}(S^n)$. One notices that the number of discretization points remains constant. In dimension more than one the connectivity remains unchanged as well. If mesh adaptation is used between two deformations the connectivity changes. Then one needs to use a transient fixed point algorithm for the set (metric, mesh, solution) in order to reduce the impact of connectivity changes on the solution (see Chapter 4).

Discrete (11.7) Finally, discrete domain deformation for $\Omega_{1h}(t)$ is obtained by solving (11.7) by P^1 finite elements. This propagates the deformation of $\Omega_{2h}(t)$ defined by U through $\Omega_{1h}(t)$.

To summarize, the following steps have to be taken by this coupling (below we will see more sophisticated coupling strategies):

- knowing $U^n(S^n)$;
- solve (11.7): $\Omega_{1h}(t^n) \to \Omega_{1h}(t^{n+1})$ by $y^{n+1} = y^n + Y^{n+1}$;
- solve (11.9) and get u^{n+1};
- using $u^{n+1}(S^n)$, solve (11.11) and get a new U^{n+1}.

11.4.1.2 Equivalent boundary condition Small variations around $S = 0$ can be expressed by changing the boundary condition at the interface between models keeping the domain unchanged. This has the obvious advantage of bringing serious simplification to the coupling algorithm (see also Chapter 6). The overall model now reads:

$$u_t + (cu)_y - (\mu u_y)_y = f(y,t), \quad \text{on }]-1,0[\tag{11.13}$$

$$U_{tt} - U_{yy} = u(0,t)\delta(0), \quad \text{on }]0,1[\tag{11.14}$$

with initial and boundary conditions given by:

$$u(y,t) = u_0(y) \quad \text{on }]-1,0[,$$

$$U(y,t) = U_0(y), \; U_t(y,0) = U_1(y) \quad \text{on }]0,1[,$$

$$u(-1,t) = 0, \; U(1,t) = 0,$$

the equivalent boundary conditions at the interface $S = 0$ are derived noticing that $S(t) = U(0,t)$:

$$u_y(0,t) = u_y(S(t),t) - u_{yy}(0,t)S(t) = -u_{yy}(0,t)U(0,t),$$

$$U_y(0,t) = U_y(S(t),t) - U_{yy}(0,t)S(t) = -U_{yy}(0,t)U(0,t).$$

11.4.2 Coupling strategies

The coupling above was semi-implicit in the sense that when information is available it is used. This can bring a time lag between the two models. The following diagram gives a sketch of this coupling:

$$S : n \longrightarrow n+1 \longrightarrow n+2$$
$$\nearrow \uparrow \qquad \nearrow \uparrow$$
$$F : n \longrightarrow n+1 \longrightarrow n+2$$

For the sake of simplicity, let us show different possible coupling strategies on a first-order system:

$$\dot{Z} = f(Z), \quad Z(t=0) = Z_0. \tag{11.15}$$

A second order dynamic system can be put in first-order form using an extra variable for the derivative. Z gathers the fluid and structure variables. The governing equations are therefore the fluid and structure models with terms gathered in $f(Z)$. In a control problem this system also includes control parameters as one also has a dynamical system for the evolution of the control parameters (see Chapter 7).

The easiest way to couple several models is a parallel approach as in the following first-order explicit scheme where both models are advanced in time and the necessary information is communicated from one model to the other:

$$Z_0 = Z(t=0), \quad Z^{n+1} = Z^n + \Delta t f(Z^n). \tag{11.16}$$

There is no time lag between models as shown in the following diagram:

$$S : n \to n+1 \to n+2$$
$$\updownarrow \qquad \updownarrow$$
$$F : n \to n+1 \to n+2$$

Obviously, the time interval between steps n and n+1 is not necessarily the time steps used in the calculation. For instance, if an explicit flow solver is coupled with an implicit structural code, one will take several hundreds of fluid time iterations for one iteration of the structure model. In particular, no communication is made between the fluid and structure models during the intermediate flow iterations leaving unchanged the information transfer and dependency between models. This is important as one would like minimum communication between the different models. Another popular choice is to use a pseudo-time-stepping approach discussed in Chapter 12 for the coupled solution of microfluidic flows. Iteration $n+1$ from n is obtained as a steady solution in τ of:

$$Y_\tau + \frac{Y - Z^n}{\Delta t} = f(Y), \quad Y(0) = Z^n,$$

introducing subcycling in τ, the following iterations in p:

$$\frac{Y^{p+1} - Y^p}{\Delta \tau} + \frac{Y^p - Z^n}{\Delta t} = f(Y^p), \quad Y^0 = Z^n,$$

corresponds to fully implicit iterations in n when $\|Y^{p+1} - Y^p\| \to 0$:

$$\frac{Z^{n+1} - Z^n}{\Delta t} = f(Z^{n+1}), \quad Z^{n+1} = Y^p.$$

Without being exhaustive, to improve time accuracy one can use either multiple sub-iterations with Runge Kutta methods or higher order integration rules such as the trapezoidal rule:

$$Z_0 = Z(t=0), \quad Z^{n+1} = Z^n + \Delta t \frac{f(Z^n) + f(Z^{n+1})}{2}. \tag{11.17}$$

In practice a fully implicit scheme is difficult to implement. One can use an approximation or prediction of Z^{n+1} by:

$$Z^{n+1/2} = 2Z^n - Z^{n-1} = Z^n + \dot{Z}^n \Delta t.$$

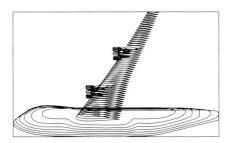

FIG. 11.5. Structural model and surface CFD mesh.

11.4.3 *Low-complexity structure models*

Let us describe now some of the models one can consider for the elastic behavior of the shape. One would like to consider simple models that are easy to include in the design process as our motivation is to always perform cheap design and only a posteriori validate with the full machinery.

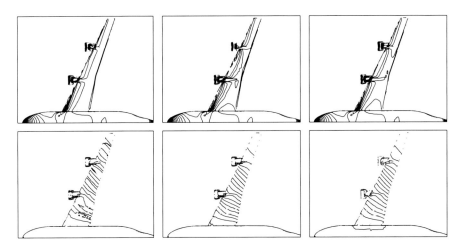

FIG. 11.6. Iso-normal deformations (upper) to the shape and iso-Mach contours for a coupled inviscid simulation. One sees that iso-contours of structural deformations are orthogonal to those of Mach contours. This is why meshes for fluid and structure models are often incompatible.

More precisely, one would like to cheaply include elastic requirements during the aerodynamic design and inversely enrich the structural design using low-complexity fluid models. Usually the fluid and structure designs are performed separately, based on iterative projection schemes: one optimizes the fluid functionals trying to keep unchanged the structural characteristics, and inversely. As an example, one can be asked for aerodynamic shape optimization under thickness constraints. The surface shape is then the parameter for fluid optimization and the local thickness of the structure is the parameter in structural optimization. Figure 11.7 shows an example of such optimization where the local thickness and the surface shape are optimized under mutual aerodynamic and structural constraints.

This is linked to multi-criteria and model optimization issues and one should use partition of territories and game theory for a suitable definition of the design parameters for each field in order to minimize the effects of incompatible requirements on the design [12].

Suppose we have a set of parameters $X = \{x, y\}$ which we split into two parts. The splitting is after one has identified the subset of parameters active for the flow and for the structure behavior. One somehow define two territories. Now the optimization of two functionals $J_1(X)$ and $J_2(X)$ describing some flow and structure features can be based on building a sequence of ($X_n^* = \{x_n^*, y_n^*\}$) solutions of:

$$x_{n+1}^* = \mathrm{argmin}_x J_1(x, y_n^*), \quad \text{such that} \quad J_2(x_{n+1}^*, y_n^*) \leq J_2(x_n^*, y_n^*),$$

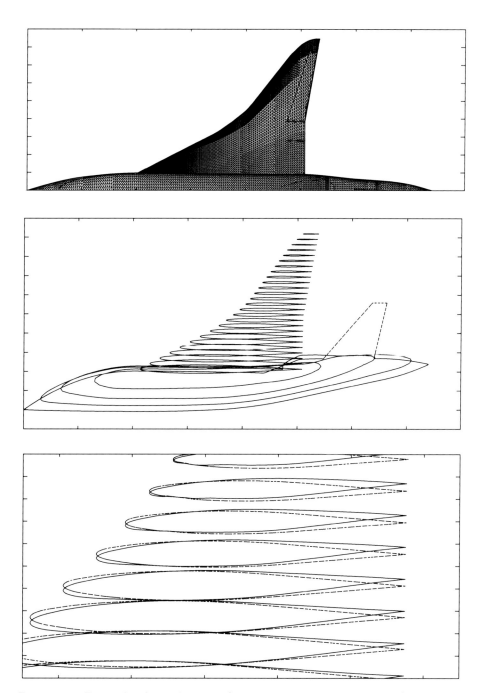

FIG. 11.7. Example of aerodynamic (drag minimization at given lift) shape and thickness optimization. Upper; the CAD-free parameterization for the aerodynamic optimization (the surface mesh). Middle: parameters for structural thickness optimization. The initial by-section definition is made from the CAD-free parameterization. Below: the initial and final shapes. The maximum thickness has been conserved for each section.

$$y_{n+1}^* = \text{argmin}_y J_2(x_n^*, y), \text{ such that } J_1(x_n^*, y_{n+1}^*) \leq J_1(x_n^*, y_n^*).$$

If this sequence converges the solution means that no better solution can be found minimizing each subproblem alone using the available information from the other subproblem. This is called a Nash equilibrium [8] where *no unilateral deviation in strategy by any single player is profitable.*

From a practical point of view, the splitting is often motivated by incompatible requirements (see the iso-contours in Fig. 11.6). This is sometimes because the fluid and structure solvers are in the hands of different people and a different team is in charge of the whole simulation and design chain (the situation gets worse adding extra specifications on electromagnetism or noise, for instance). Therefore, if one can account through low-complexity modeling for the main requirements of both fields, the final design should better suit both fields.

The first elastic model one can consider for an object might be quite simple only having three degrees of freedom $X = (\alpha_x, \alpha_y, \alpha_z)$ based on the three Euler angles. The variations of these angles can be described by three coupled ordinary differential equations:

$$I\ddot{X} + \Lambda \dot{X} + KX = S(X, \dot{X}, C_l, C_m, C_d), \tag{11.18}$$

where I, Λ and K describe the structure characteristics. S contains the aerodynamic forces acting on the structure. S also contains X and $\dot{X}$ to enforce the coupling between the flow and structure systems. For a 3D wing, for instance, having its chord along the x direction and its span along y, the most important displacement is due to the lift coefficient acting on α_x, then due to the aerodynamical moment acting on the pitching angle α_y and finally due to the drag forces through α_z. Linearizing (11.18), we have:

$$I\overline{\ddot{\delta X}} + \left(\Lambda - \frac{\partial S}{\partial \dot{X}}\right) \overline{\dot{\delta X}} + \left(K - \frac{\partial S}{\partial X}\right) \overline{\delta X} = 0,$$

which can have undumped modes following the sign of $\partial S / \partial \dot{X}$.

A more sophisticated model can be based on ODEs for eigenmodes of the structure. This approach is widely used in industry. It is based on the assumption that the structural modes are weakly modified by the flow (example of coupling in Figs. 11.5 and 11.6). This assumption appears valid in most situations [10].

One can also build surface-based elastic models and enrich a CAD-free parameterization with elastic features. Our aim again is to have only one geometrical entity during simulation and design and the same also when doing MDO configurations. The geometrical definition is therefore again the surface discretization (in 2D segments and in 3D triangles as presented in Chapter 5). The definition of normals to the surface is therefore an easy task.

Using an analogy with the models described above one can introduce an elastic model for the displacements of the shape in the normal direction (the normal deformation between instant t^n and t^{n+1} can be defined as $X = \int_{t^n}^{t^{n+1}} \vec{x}.\vec{n} dt)$ introducing a partial differential equation with first and second-order time derivatives and with second and fourth order elliptic operators for instance:

$$M\frac{\partial^2 X}{\partial t^2} + C\frac{\partial X}{\partial t} - K\Delta X - \mu\Delta^2 X = F. \qquad (11.19)$$

The first two terms involve point-wise behavior of the shape and the third and fourth terms link the surface nodes together. The parameters M (mass), C (damping), K (stiffness) and μ (shell stiffness) encapsulate the mechanical characteristics of the structure and should be identified in order to reproduce the overall behavior of the original structure. The presence of second and fourth-order surface space derivatives enables the model to produce both membrane and shell-type behavior. F gathers the aerodynamic forces on the structure.

This subject has been widely investigated considering two or more solvers with incompatible interface discretizations in distributed environments [1, 2, 9, 11]. With incompatible parameterizations for the fluid and structure interface, a major difficulty comes from information transfer between the two models.

11.5 Stability, robustness, and unsteadiness

We saw that for time-dependent optimization problems functionals can be formulated in the frequency domain. Consider a dynamical system

$$u_t + F(u) = 0, u(p, t = 0) = u_0 \qquad (11.20)$$

with $u = u(p,t)$ the state and p a parameter. It is indeed natural to consider this system stable around an equilibrium state $u(p)$ (i.e. $u_t = 0$) if the eigenvalues λ of the linearized system

$$F_p(u(p))P = -F(u(p)) = 0$$

have no negative real part.

Considering the lowest frequency of the system, this can be formulated as:

$$\text{Re}(\lambda_{\min}) \geq 0,$$

with $\lambda_{\min}$ the eigenvalue of $F_p(u(p))$ with the smallest real part.

In the context of fluid-structure coupling, to avoid resonance, one must ask in addition for the lowest fluid frequency $\lambda_{\min}^f$ to be higher than the lowest structural frequency $\lambda_{\min}^s$:

$$\text{Re}(\lambda_{\min}^f) \geq \text{Re}(\lambda_{\min}^s).$$

If a modal representation is used for the structure, $\text{Re}(\lambda_{\min}^s)$ known. As one usually assumes the modes of the structure remain unchanged during the coupling, this constraint can be expressed using information from the fluid side alone.

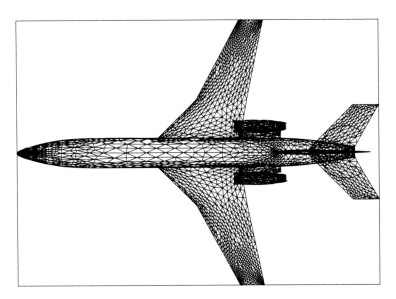

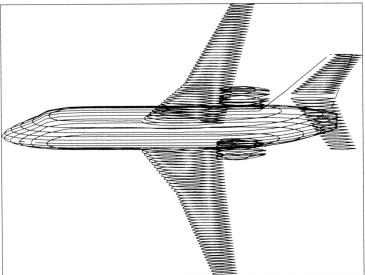

FIG. 11.8. Aeroelastic simulation and control for a business jet. Fluid surface (upper) mesh and structural model (lower). The control aims at keeping bounded structural oscillations in wings using small instantaneous incidence correction through flap deflection. The control is defined by (11.21) using incomplete and instantaneous sensitivity evaluation for the instantaneous lift coefficient. To make the control realistic the amplitude of incidence corrections has been bounded by $1°$. Figure 11.9 shows the wing tip oscillations with and without the control.

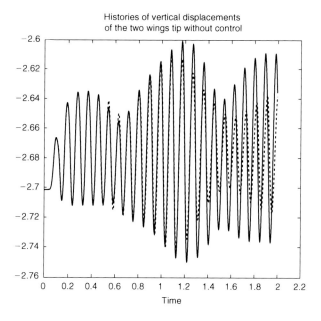

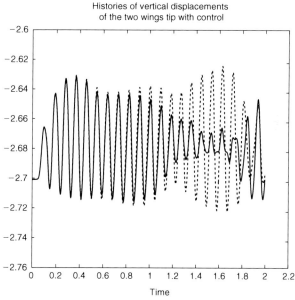

FIG. 11.9. Aeroelastic simulation and control for a business jet. Wing tip vertical evolution in time without (upper) and with (lower) control. This is a sub-optimal control as incomplete sensitivity is used and also because the maximum incidence correction is bounded by 1° and the control frequency by 2 Hz.

One natural way to apply this idea to shape optimization is to ask for the optimum to realize similar stability requirements evaluated for the Hessian of the functional with $F(u(p)) = \nabla_p J(u(p))$. Indeed, for robust optimization one aims at an optimum having:

$$\nabla_p J(u(p)) = 0, \quad \text{and} \quad \text{Re}(\lambda_{\min}(\nabla_{pp} J)) \geq 0,$$

which is naturally achieved in the context of convex optimization. In practice, the Hessian is not available. If one uses a BFGS formula (see Chapter 7) producing a symmetric positive definite approximation of the Hessian the quadratic approximation of the problem is therefore stable and so is the original problem if the BFGS approximation is valid. If this eigenvalue has multiplicity one it can be estimated by a power iteration method for the inverse of the Hessian approximation as the largest eigenvalue [11].

Now, accounting for unsteadiness can also be seen as growth of small perturbations in the system. Let us consider the linearization in the state of (11.20):

$$v_t + F_u(u(p))v = 0, \quad v(0) = \varepsilon.$$

If the flow is unsteady F_u has eigenvalues with negative real parts at $u(p)$, and in particular:

$$\text{Re}(\lambda_{\min}(F_u(u(p)))) < 0.$$

Therefore, another alternative to functionals such as (11.3) or (11.4) for shape optimization with constraint on unsteadiness control is to require for $\text{Re}(\lambda_{\min}(p)) = \text{Re}(\lambda_{\min}(F_u(u(p))))$ to increase during optimization:

$$\text{Re}(\lambda_{\min}(p)) \leq \text{Re}(\lambda_{\min}(p_{\text{opt}})).$$

11.6 Control of aeroelastic instabilities

The problem of interest is to predict the aeroelastic behavior of the coupled system and to control situations where the structure behavior becomes unstable due to fluid perturbations. Control has been performed for all cases using the body incidence. This is easy to prescribe for an on-board device acting on the flaps.

We consider the flow over a business jet. The fluid and structure systems have been excited by inflow periodic perturbations. The perturbation frequency is 10 Hz and its amplitude is less than $1°$ for both inflow incidence angles. These perturbations might happen in stormy weather or during take-off and landing in the wake of bigger aircraft. One sees that a control law based on small incidence

---

[11] If the largest eigenvalue in modulus of a matrix A is of multiplicity one, it can be obtained by $V^{k+1} = \frac{AV^k}{\|AV^k\|}$, $\lambda_{\max} = \lim_{k \to \infty} \frac{(AV^k, V^k)}{(V^k, V^k)}$.

corrections defined by instantaneous incomplete sensitivity of lift fluctuations is effective:

$$\delta\alpha_i = -\rho\nabla_{\alpha_i}(\partial_t C_l(t))^2, \quad i = 1, 2 \tag{11.21}$$

where α_i are the three aircraft incidences angles. These corrections are bounded by $1°$. This control behaves like a Kalman filter in the context of nonlinear coupling (see Fig. 11.8 and 11.9). The control is adaptive in the sense that if the flow conditions change the control naturally adapts to the new flow.

References

[1] Donea, J. (1982). An ALE finite element method for transient fluid-structure interactions, *Comp. Meth. App. Mech. Eng.* **33**, 689-723.

[2] Farhat, C. and Lesoinne, M. (1996). On the accuracy, stability and performance of the solution of three-dimensional nonlinear transient aeroelastic problems by partitioned procedures, *American Institute of Aeronautics and Astronautics*, **96-1388**.

[3] Leclerc, E. Mohammadi, B. and Sagaut, P. (2007). On the use of incomplete sensitivities for feedback control of laminar vortex shedding, *Comput. Fluids*, **35(10)**, 14321443.

[4] Medic, G. Mohammadi, B. and Stanciu, M. (1998). Prediction and aeroelastic simulation of turbulent flows in civil engineering applications, *Proc. Eccomas 98*, Athenes.

[5] Medic, G. Mohammadi, B. Petruzzelli, N. and M. Stanciu. (1999). 3D optimal shape design for complex flows: application to turbomachinery, *American Institute of Aeronautics and Astronautics*, **99-0833**.

[6] Medic, G. Mohammadi, B. Petruzzelli, N. and M. Stanciu. (1999). Optimal control of 3D turbulent flow over circular cylinder, *American Institute of Aeronautics and Astronautics*, **99-3654**.

[7] Mohammadi, B. (1999). Dynamical approaches and incomplete gradients for shape optimization and flow control, *American Institute of Aeronautics and Astronautics*, **99-3374**.

[8] Nash, J. (1950). Equilibrium points in n-person games. *Proceedings of the National Academy of Sciences*, **36(1)**, 48-49.

[9] Nkonga, B. and Guillard, H. (1994). Godunoc type method on non-structured meshes for three dimensional moving boundary-problems, *Comp. Meth. Appl. Mech. Eng.* **113**, 183-204.

[10] Ohayon, R. and Morand, J.P. (1995). *Fluid Structure Interaction*, John Wiley, London.

[11] Piperno, S. Farhat, C. and Larrouturou, B. (1995). Partitioned procedures for the transient solution of coupled aeroelastic problems, *Comp. Meth. Appl. Mech. Eng.* **124**, 97-101.

[12] Tang, Z. Desideri, J.-A. and Periaux, J. (2002). Distributed optimization using virtual and real game strategies for aerodynamic design, *INRIA-RR*, **4543**.

12
FROM AIRPLANE DESIGN TO MICROFLUIDICS

12.1 Introduction

In this chapter we show the application of optimal shape design to microfluidic flows. Microfluidics is a recent field of research. It deals with flows in structures bringing into play dimensions characteristic of the order of a micrometer. These flows develop nanometric characteristic lengths, for instance in boundary layers, and the range of variation of variables is multi-scaled which makes the calculation of these flows complex. The modeling often involves multi-physics. The behavior of the system is governed by several coupled variables. Sometimes, the quantities are not detectable by the available measuring apparatus and modeling and simulation are valuable tools to develop approaches to make them detectable.

Integrated electrokinetic microsystems have been developed with a variety of functionalities including sample pretreatment, mixing, and separation. Electric fields are used either to generate bulk fluid motion (electroosmosis) or to separate charged species (electrophoresis) [1, 2, 7, 9, 13, 14]. Examples of separation assays include on-chip capillary zones, electrophoresis and isoelectric focusing. Preconcentration methods include field amplified sample stacking and isotachophoretic preconcentration. These applications involve the convective-diffusion-electromigration of multiple ionic and neutral species.

Again we shall try to introduce low-complexity modeling and incomplete sensitivity for both cheap simulation and design and also a better understanding of the mechanisms involved.

Our experience is based on the prototyping of designed devices performed by Prof. Juan Santiago at the Microfluidic Laboratory at Stanford University.

The applications of microfluidic devices are multiple. Let us quote some examples:

- Health: biomedical research, DNA sequencing, protein folding, detection of possible chemical reactions with other substances (used for instance for tests of new molecules on a given target), rapid mixing.
- Aeronautics and space: switches, actuators and microfluidic measurement instruments.
- Environmental applications: detection of pollution in water or food.
- Data processing: cooling of the computers by pumps not having moving parts.

Usually, the realization of these devices uses the techniques of micro-electronic engraving, mainly on silicon supports. The optical techniques require dynamic microscopy and the use of neutral substances excitable by certain frequencies of light and fluorescence. We notice that solutions discovered after modeling, simulation and optimization are often nonintuitive. In other words, mathematical modeling allowed in certain cases a return to physical modeling, in particular by identifying the significant points in a coupled situation. Thus the approach allows for the development of subsets with reduced complexity which can be used in efficient design platforms.

In this chapter we consider the following problems:

- Optimization of the extraction of microfluidic samples before their introduction in transport for separation or stacking devices.
- Shape optimization in microfluidic devices.
- Modeling and optimization of rapid mixing devices used for protein folding.

12.2 Governing equations for microfluids

The system of equations applicable to electrokinetic microfluidics usually gathers a Stokes system and advection-diffusion equations for the species with a coupling between the two sets of equations through the electric field. This is briefly described in Section 12.7. The Stokes system is suitable because for typical electrokinetic microchannel applications the observed flow motion has a velocity of about $10^{-4}\ m/s - 10^{-3}\ m/s$, channel thickness of 100 μm and kinematic viscosity of about $10^{-5}\ m^2 s^{-1}$. This leads to Reynolds numbers ranging from 0.001 to 0.01.

Also, due to spontaneous charge separation that occurs at the channel walls there is formation of an electric double layer [13]. The typical size of this layer is a few nanometers. The stiffness of this electric double layer makes it difficult to compute using classical numerical approaches applied to the Stokes model. However, it is known that at the edge of the double layer the flow is parallel and directly proportional to the electric field [7, 13]:

$$u = -\zeta\ \nabla \phi. \tag{12.1}$$

Here ζ is a positive constant which depends on the material with which channels are built and involves what is called zeta potential and also the fluid permittivity and dynamic viscosity. The flow is therefore parallel to the wall and to the electric field. We will use this information to derive boundary conditions and low-complexity models for the flow. This is similar to the derivation of wall functions for turbulent flows.

12.3 Stacking

Let us describe an important phenomenon in microfluidic flows and introduce an example of low-complexity modeling. The aim of stacking is to amplify the local concentrations of species to make them observable by existing devices. One uses the fact that an ion sample passing from a region of high to another of low conductivity will accumulate at the interface between the regions. An analogy can be given with particles in a supersonic region placed before another subsonic region. In the subsonic region the flow velocity is lower and therefore particles in the supersonic region will catch those subsonics.

Consider the case of three ionic species. The third species is the sample species of interest whose molar concentration is low and one would like to increase it through stacking. Typically the sample species is three or four orders of magnitude smaller than other species and consequently one needs to stack the sample species by a factor of about 1000 to make it observable. Representative experimental conditions are described in [1, 9, 12].

A single interface of conductivities is represented by a simple error function profile for the initial concentration:

$$c^1_{init}(x) = \tfrac{1}{2} b_1 \left(\gamma + 1 - (\gamma - 1)\text{erf}(\alpha x)\right)$$
$$c^3_{init}(x) = b_3 \left(1 + \text{erf}(\alpha x)\right) \quad (12.2)$$
$$z_2 c^2_{init}(x) = -z_1 c^1_{init}(x) - z_3 c^3_{init}(x)$$

where $b_1 = 100$, $b_3 = 0.1$, $\alpha > 0$ and $\gamma > 0$. Note that α controls the sharpness in the initial plug and γ indicates the initial concentration ratio between high and low conductivity regions in the sense that

$$\lim_{x \to -\infty} c^1_{init}(x)/b_1 = \gamma, \quad \lim_{x \to \infty} c^1_{init}(x)/b_1 = 1.$$

An important property of γ, both experimentally and numerically, is that it also indicates the final stacking ratio of c^3. Let us check that on a one-dimensional model. Neglecting diffusion and the motion of the ambient fluid, the species conservation equation for different species becomes

$$c^i_t + \frac{\partial}{\partial x}\left(z_i \nu_i c^i \phi_x\right) = 0. \quad (12.3)$$

A constant current density j is applied in the axial direction defined by the difference of the electric potential at the ends of the channel:

$$j = \frac{V_{\text{left}} - V_{\text{right}}}{L} \sigma_{\text{total}},$$

where L is the length of the channel and

$$\sigma_{\text{total}} = \int_0^L \sigma(x,t)dx, \quad \sigma(x,t) = \sum_i z_i^2 \nu_i c^i.$$

Let us further suppose that σ_{total} is independent of time. The electric field is then locally defined as:

$$E = \phi_x = \frac{j}{\sigma}. \quad (12.4)$$

Now in our setting of three species the concentration of sample species c^3 increases as it migrates from a region of high conductivity to a region of lower drift velocity. Hence the stacked sample keeps increasing with time and the concentration progressively approaches a maximum steady value. Now in the steady state the net flux of the species c^3 at the left and right edges balances:

$$(z_3\nu_3 c^3 E)|_{\text{left}} = (z_3\nu_3 c^3 E)|_{\text{right}}.$$

Then considering the initial conditions (12.2) we have

$$\sigma_{\text{left}} = \sum_{i=1}^{3} z_i^2 \nu_i c_{\text{left}}^i \approx z_1^2 \nu_1 c_{\text{left}}^1 + z_2^2 \nu_2 c_{\text{left}}^2 \approx (z_1^2 \nu_1 - z_1 z_2 \nu_2) b_1 \gamma.$$

As j is constant this implies

$$\frac{c_{\text{left}}^3}{c_{\text{right}}^3} = \frac{E_{\text{right}}}{E_{\text{left}}} = \frac{\sigma_{\text{left}}}{\sigma_{\text{right}}} \approx \gamma.$$

When diffusion and motion of the ambient fluid are present the above arguments can still be used to provide an upper bound for the stacking capacity. Indeed, both diffusion and the convective motion tend to distort the front where stacking happens and hence one expects a lower final stacking ratio in this case.

Figure 12.1 shows the behavior of a one-dimensional model made from equation (12.4) for the electric field, an algebraic equation such as (12.1) linking the electric field and the flow velocity (no pressure effects) and equations (12.3) for the different species. The model produces the correct stacking limit and migration time. Despite its simplicity, this model is therefore suitable as a low-complexity model in design. For instance, it gives enough information for the definition of the length of a separation and stacking channel. It also gives good estimations of the different species migration times. It can therefore be used to characterize the initial plug characteristics. Below, we will see other low-complexity models in shape design.

12.4 Control of the extraction of infinitesimal quantities

The first control problem we consider is a state control problem using the applied electric fields through the potentials V_n, V_s, V_{e1}, V_{e2} and V_w shown in Fig. 12.2). Indeed, before any analysis, it is a question of being able to extract a suitable sample from the solution to be analyzed. This sample must have certain geometrical characteristics with in particular very low dispersion and with the concentration gradient parallel to the electric field, which means iso-contours of concentration normal to the walls (see Fig. 12.3 [1, 10]). This is essential for the quality of the analysis as detection tools are highly sensitive to the dispersion of the sample. Separation occurs during transport due to the difference of mobility of species in the electric field.

12.5 Design of microfluidic channels

Separation improves with the length of the channel and the intensity of the electric field [3, 11]. This former quantity cannot exceed some limit for stability purposes [9, 14] and the required compactness of the devices (10 meter long

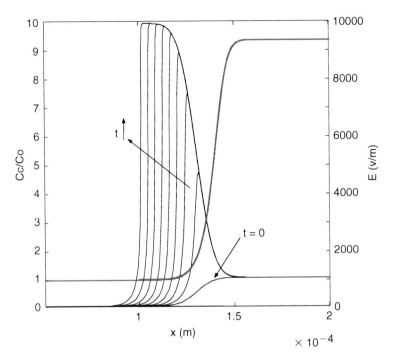

FIG. 12.1. Low-complexity model reproducing the right theoretical stacking ratio limit (here $\gamma = 10$).

channel on a support of 10 cm^2) implies the introduction of turns which in turn have undesirable effects as they introduce skew in the advected bands. Skew implies a dispersion of the electrophoretic sample bands in the flow. This leads to a shape optimization problem. By skew we mean that if iso-values of species concentration are normal to the wall before a turn, they lose this property after (see Fig. 12.6). This curved-channel dispersion has been identified as an important factor in the decrease of separation efficiency of electrophoretic microchannel systems. Unfortunately, we notice that reducing the skew often introduces a new type of residual dispersion associated with band advection away from the channel boundaries. We also notice that to avoid this effect it is necessary for the channel walls to be as smooth as possible with minimal curvature variation (see Fig. 12.7). This is somewhat contradictory to the shapes obtained from a minimization based only on skew minimization. Regularity of the shape needs to be controlled as well during optimization.

The optimization formulation for such devices has to include therefore the following points:

- minimize the skew due to turns;
- minimize the residual dispersion associated with band advection;

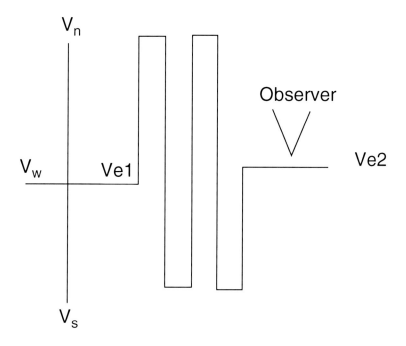

FIG. 12.2. Typical geometry of an extraction/separation device. V_n, V_s, V_{e1}, V_{e2} and V_w denote the different electric potentials applied.

- minimize variations in walls curvature;
- maximize the length of the channel;
- minimize the overall size of the device.

The skew can be qualified in different ways. For example, we can ask for isovalues of the advected species to be always normal to the flow field and consider a functional of the form:

$$J(x) = \int_0^T \int_\Omega (\nabla c(x,t) \times u(x))^2 dx dt, \qquad (12.5)$$

where T is the total migration time. This functional is not suitable for sensitivity evaluation using incomplete sensitivity concepts described in Chapter 8 as it involves information over the whole domain. In addition, this cost function is too restrictive as we are actually interested only in minimizing the final skew (i.e. after the turn) and therefore consider, for instance:

$$J(x) = \int_\Omega (\nabla c(x,T) \times u(x))^2 dx. \qquad (12.6)$$

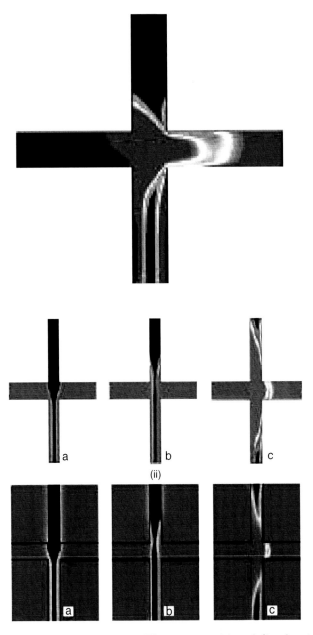

FIG. 12.3. Upper: initial extraction. The extracted band (in the right channel) will be transported for separation. In this example the initial band is not suitable for transport. We aim at a band with minimum dispersion and with iso-contours orthogonal to the walls. Middle: after optimization. The control law has been experimentally validated (lower). Control uses pinching and pull-back steps using applied electric fields.

This functional is still over the whole domain. Another way to define the skew which avoids the previous difficulty is to ask for all particles traveling on characteristics to have the same migration time.

We therefore consider:
$$J(x) = \left(\int_\chi \frac{ds}{\|u\|} - \int_{\chi'} \frac{ds}{\|u\|} \right)^2, \quad (12.7)$$

for any couple of characteristics χ and χ' linking the outlet to the inlet. Here again, the cost function is over the whole space, but we can consider only a few characteristics. The two main characteristics are those defined by the internal and external walls of the channel:
$$J(x) = \left(\int_{\Gamma_i} \frac{ds}{\|u\|} - \int_{\Gamma_o} \frac{ds}{\|u\|} \right)^2, \quad (12.8)$$

where Γ_i is the inner wall and Γ_o the outer wall in a turn. We recall that in our model velocity is parallel to the walls from (12.1). This last formulation is interesting as it only involves boundaries which we know to be suitable for incomplete sensitivities. Another interesting feature of formulations (12.7) and (12.8) over (12.5) is that they do not require knowledge of the distribution of the advected species. Also one notices that dispersion increases with the variation of the shape curvature (see Fig. 12.7). One therefore tries to find variations bringing the least modification in curvature over simple 90° and 180° curves. These two curves are sufficient to build any channel pattern. Variations of curvature along inner and outer channel walls are represented through:
$$\left(\int_{\Gamma_i} \left\| \frac{\partial n}{\partial s} \right\| - \int_{\Gamma_i^0} \left\| \frac{\partial n}{\partial s} \right\| \right)^2 + \left(\int_{\Gamma_o} \left\| \frac{\partial n}{\partial s} \right\| - \int_{\Gamma_o^0} \left\| \frac{\partial n}{\partial s} \right\| \right)^2, \quad (12.9)$$

where s is the curvilinear abscissa and 0 denotes initial inner and outer walls. Minimization should be performed keeping this constrained close to zero. To achieve compactness of the global device, it is preferable not to move the external walls, especially in 180° turns. This is satisfactory as then the second term in expression (12.9) vanishes. Figure 12.4 shows various shapes achived for a 90° turn with different regularity requirements.

Cost function (12.8) is suitable for the application of incomplete sensitivities. We can however increase direct geometrical contributions by the fact that the velocity is parallel to the walls. The cost function we consider for derivation with respect to the walls is therefore:
$$J(x) = \left(\int_{\Gamma_i} \frac{ds}{\tau.u} - \int_{\Gamma_o} \frac{ds}{\tau.u} \right)^2. \quad (12.10)$$

where τ is the local unit tangent vector to the wall. In incomplete sensitivity, u is frozen when linearizing (12.10). We compare this approximation of the gradient with the full gradient by finite differences (see Fig. 12.5).

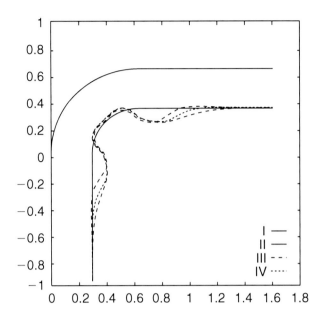

FIG. 12.4. Shapes obtained under the same optimization conditions for three admissible spaces with different minimum regularity required for the shape (I: initial, II, III, IV: increasing regularity required).

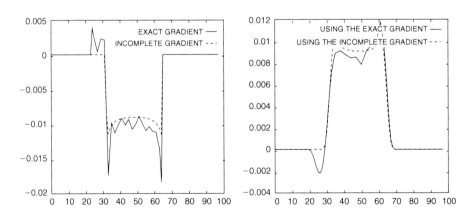

FIG. 12.5. Sensitivity evaluation around the initial 90° turn for control points along the inner channel wall. Comparison between the exact and incomplete sensitivities (left). Right: the deformations obtained using these gradients.

12.5.1 Reduced models for the flow

We take the opportunity of this shape optimization to again show how to avoid considering the full physical model during optimization. The governing equations have been described in (12.14) and (12.15) in the electroneutral case. We saw how to reduce the full system for one-dimensional problems to quantify the optimal length of channels needed for separation. We saw how to simplify the functional to bring it into the domain of validity of incomplete sensitivities. This is similar to making the following hypothesis for the state equations.

The calculation of flow velocity and migration of species have been uncoupled. In other words, in system (12.14) the species are considered as passive. The pressure gradient is supposed to be negligible and the velocity field is everywhere proportional to the local electric field and therefore parallel to the walls following expression (12.1).

$$u \parallel E = -\nabla \phi. \tag{12.11}$$

Indeed, in the design problem of interest it is enough for the velocity vector to be known up to a multiplicative constant: we want the migration times along the two walls to be the same, but the exact value of this time is of no interest for the design.

Once the design is made using cost function (12.8), we need to check that the final design indeed reduces the skew. This can be checked by advecting a plug for a single species c by the velocity field defined above. Again, the speed of advection is of no interest. Most important the velocity field is a stationary field. Advection is time dependent but it is only a post-processing step.

We show the skews produced by a 180° turn in Fig. 12.6. No symmetry assumption has been made during design. Figure 12.7 shows three classes of optimal shapes for the 180° turn. The optimal shape has been prototyped and experimental results correspond to the predicted behavior by the low-complexity model described above [11].

12.6 Microfluidic mixing device for protein folding

Another application of interest is to build mixing devices for protein folding applications [5, 8]. These devices are used in biomedical research. One looks for different configurations in a ribosome chain having different behavior in the same situation. For instance, one might be interested in finding a protein reacting to the presence of a given virus. The optimization aims to control the folding time and in particular to reduce it to force the chain folding in other ways than its natural tendency. All other parameters being frozen, the final shape of the chain depends on the time it takes to fold. Up to now the largest limitation concerns the treatment of times shorter than the molecular diffusion characteristic time. This time is proportional to the square of the length-scale which implies a constraint on the size of the device. One would like to use other mechanisms involved than diffusion to reduce the folding time. In particular, we aim to take advantage of transport and pressure effects.

FIG. 12.6. Concentrations for the initial 180° turn: effect of the turn on the advected species.

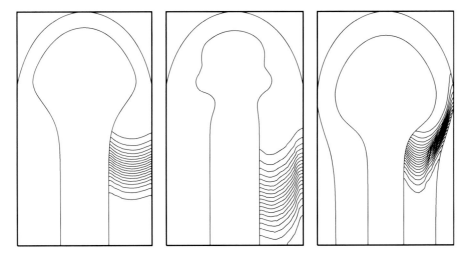

FIG. 12.7. Three optimal shapes for the 180° turn. The two shapes on the right are local minima.

The mixer shape considered is a typical three-inlet/single-outlet channel architecture. The model is symmetric so we only study half of the mixer [6]. Our model is a 2D approximation of the physical system [4]. Experiments show a 5% deviation from a 3D model which is satisfactory for a 2D model to be used as a a low-complexity model in optimization [6].

Our aim is to optimize the corner shapes (see Fig. 12.9). We parameterize the corner regions by cubic splines. The total number of parameters is eight, four for each corner. The total dimension of the micro-mixer cannot exceed $22\mu m$ long and $10\mu m$ large. The lithography step in fabrication limits the minimum feature size to a minimum of 1 μm (i.e. the minimum channel thickness one can realize). The width of the side channel nozzles is set to 3 μm and the width of the center channel nozzles to 2 μm. The maximum side velocity is $U_s = 10^{-4} m/s$. A typical flow Reynolds number based on side channel thickness and the flow inlet is about 15.

The cost function to minimize is the mixing time of the considered Lagrangian fluid particle traveling along the centerline. The mixing time is defined as the time required to change the local concentration from 90% to 30% of the inlet value c_0:

$$J(x) = \int_{y_{90}}^{y_{30}} \frac{dy}{\|u\|}, \qquad (12.12)$$

where y_{90} and y_{30} denote respectively the points along the symmetry line where the concentration is at 90% and 30% of c_0. v is the normal component of the velocity. This cost function is not suitable for incomplete sensitivity as it is not a boundary integral and it does not involve the product of states by geometry entities. One therefore needs to compute the full gradient. However, because we want to use a black-box solver for the state solution one cannot use the adjoint approach. One rather uses a multi-level evaluation of the gradient as describe in Chapter 8.

More precisely, we use a different level of discretization for the state u with the same state equation looking for state sensitivity on a coarse mesh while the state is evaluated on a much finer mesh (the direct loop being $x \to u \to J$):

$$d = \frac{\partial J}{\partial x}(u_f) + \frac{\partial J}{\partial u}(u_f) \cdot I_{c/f}\left(\frac{\partial u}{\partial x}(I_{f/c}(u_f))\right)$$

where subscripts f and c denote fine and coarse meshes, u_f denote the state equation solution evaluated on f, and $I_{f/c}(.)$ (resp. $I_{c/f}(.)$) is an interpolation operator between the fine and coarse meshes (resp. coarse and fine). By fine mesh we mean a mesh fine enough for the solution to be mesh independent. This means that the linearization is performed on a coarse mesh for the state contribution but around an accurate state computed on a fine mesh. Obviously if the coarse mesh tends to the fine one, the approximate gradient tends to the gradient on the fine mesh. This approach works if the approximate gradient d satisfies:

$$d.\nabla J > \varepsilon > 0,$$

which is illustrated in Fig. 12.8 showing a comparison of the full gradient with this approximation. The cost of an incomplete evaluation of the gradient using finite differences on a coarse mesh is around two evaluations of the functional on a fine mesh. On each new shape the state is recalculated on a fine mesh (with about three times more points than in the coarse mesh).

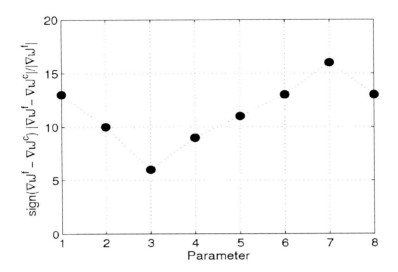

FIG. 12.8. Comparison of gradients computed by finite difference on the fine vs. coarse meshes. The sign is always correct.

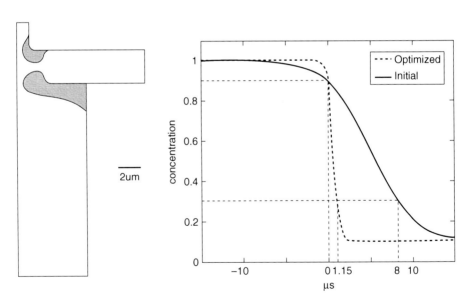

FIG. 12.9. Optimized and initial shapes (left). Concentration evolution for the initial and optimized shapes along the centerline (right).

12.7 Flow equations for microfluids

We briefly describe the governing equations for the microfluids we consider. We assume a uniform and constant electrical permittivity ε in (12.13) and low Reynolds number flows. Even though the formulation accounts for charge accumulation in the flow resulting from a coupling between electric fields and conductivity gradients, net electroneutrality is observed almost everywhere in the flow [7, 9, 13].

Let us consider the case where we have m different charged species in some ambient fluid. Let $\rho^e = F \sum_i z_i C^i = \sum_i z_i c^i$ be the net charge density where $z_i \in \mathbb{Z}$ is the valence number of species i, C^i is the molar concentration and F is the Faraday constant. The charge induces an electric field which is supposed to come from the potential ϕ. Hence we get our first equation

$$-\varepsilon \Delta \phi - \rho^e = 0, \tag{12.13}$$

where ε is the permittivity of the ambient fluid. The motion of the ambient fluid is governed by Navier-Stokes equations, but because in typical applications the Reynolds number is very low we can also use the Stokes system in the presence of an electric field:

$$\rho u_t - \mu \Delta u + \nabla p - \rho^e \nabla \phi = 0,$$
$$\nabla \cdot u = 0.$$

Here ρ is the density of the fluid and μ is the dynamic viscosity. The movement of each species is governed by $c_t^i + \nabla \cdot \mathbf{J}^i = 0$ where the current density $\mathbf{J}^i$ is given by

$$\mathbf{J}^i = -\nu_i z_i c^i \nabla \phi - d_i \nabla c^i + c^i u$$

where ν_i is the mobility times the Faraday constant and d_i the diffusivity of species i. We have now introduced all the necessary variables and parameters. Taking into account that $\nabla \cdot u = 0$ one has:

$$\nabla \cdot \mathbf{J}^i = -\nu_i z_i \nabla c^i \nabla \phi - \nu_i z_i c^i \Delta \phi - d_i \Delta c^i + u \nabla c^i.$$

This gives the system

$$\rho u_t - \mu \Delta u + \nabla p - \rho^e \nabla \phi = 0,$$
$$\nabla \cdot u = 0,$$
$$c_t^i - d_i \Delta c^i - \nu_i z_i c^i \Delta \phi - \nu_i z_i \nabla c^i \nabla \phi + u \nabla c^i = 0, \quad i = 1, \ldots, m \tag{12.14}$$
$$-\varepsilon \Delta \phi - \rho^e = 0,$$
$$\rho^e - \sum_i z_i c^i = 0.$$

The net neutrality assumption corresponds to the cases where $\rho^e = 0$. In these situations the system reduces to:

$$\rho u_t - \mu \Delta u + \nabla p = 0,$$
$$\nabla \cdot u = 0,$$
$$c_t^i - d_i \Delta c^i - \nu_i z_i c^i \Delta \phi - \nu_i z_i \nabla c^i \nabla \phi + u \nabla c^i = 0, \qquad i=1,\ldots,m \qquad (12.15)$$
$$-\Delta \phi = 0,$$
$$-\sum_i z_i c^i = 0.$$

One difficulty in solving (12.15) is that the electroneutrality constraint (the last equation) is not respected. Usually one of the species is deduced from the constraint but this usually introduces a deviation in the solution over time:

$$c^m = -\frac{1}{z_m} \sum_{i=1}^{m-1} z_i c^i. \qquad (12.16)$$

Also, in this case the physical characteristics of C^m are not taken into account during integration. Hence, it is interesting to include the electroneutrality constraint explicitly in the numerical model and use involutive completion to solve the constrained partial differential system (PDAE) [15].

These systems must be solved with the reduced order wall boundary condition (12.1) for the flow. For the electric potential we apply a Dirichlet boundary condition at the inlet and the outlet boundaries where a difference of potential is applied with homogeneous Neuman boundary condition along the walls. Together with the boundary condition for u along the wall, this condition enforces the non-penetration boundary condition for the velocity. The boundary conditions for the concentrations is Dirichlet at the inlet boundary, and homogeneous Neumann along the walls and at the outlet.

12.7.1 Coupling algorithm

We need to solve the above coupled systems cyclically. At the same time, we would like the order in the cycle not to influence the overall solution. We use the following explicit fixed point iterations. If the intermediate fixed point converges, it is fully implicit but only based on explicit iterations. Stability concerns are therefore removed and one can fix a priori time steps corresponding to the physical situation we would like to capture. However, a stability condition exists for the artificial fixed point iterations. For sake of simplicity, consider a Cauchy problem

$$y'(t) = f(y(t)), y(0) = y_0,$$

with an explicit Euler scheme $(y_{n+1} - y_n)/h_n = f(t_n, y_n)$, one needs a stability condition of the form $h_n \leq H(y_n, f(y_n))$. To overcome this difficulty we use the following algorithm:

$y_0 := y(0)$
for $n = 0 \ldots N$ do

$z_0 := y_n;\ m := -1$
repeat
    $m := m + 1$
    $(z_{m+1} - z_m)/d_m + (z_{m+1} - y_n)/h = f(t_{n+1}, z_m)$
until $|z_{m+1} - z_m| <$ tol
$y_{n+1} := z_{m+1}$
endfor

Here we have a stability condition of the form $d_m \leq G(z_m, f(y_n))$ but there is no (explicit) stability condition on h. Obviously, the accuracy in t can be improved using higher order schemes and faster convergence in m can be achieved using convergence acceleration techniques.

References

[1] Brahadwaj, R. Mohammadi, B. and Santiago, J. (2002). Design and optimization of on-chip capillary electrophoresis, *Electrophoresis J.*, **23(16)**, 27292744.

[2] Bruin, G.J.M. (2000). Recent developments in electrokinetically driven analysis of microfabricated devices, *Electrophoresis*, **21(3)**, 3931-3951.

[3] Culbeston, C.T. Jacobson, S.C. and Ramsey, J.M.. (1998). Dispersion sources for compact geometries on microchips, *Analytical Chemistry*, **70**, 3781-3789.

[4] Darnton, N. Bakajin, O. Huang, R. North, B. Tegenfeldt, J. Cox, E. Sturn, J. and Austin, M. (2001). Condensed matter, *J. Physics*, **13(21)**, 4891-4902.

[5] Hertzog, D. Ivorra, B. Babajin, O. Mohammadi, B. and Santiago, J. (2006). Fast microfluidic mixers for studying protein folding kinetics, *Analytical chemistry*, **78**, 4299-4306.

[6] Hertzog, D.E. Michalet, X. Jager, M. Kong, X. Santiago, J. Weiss and S. Bakajin, O. (2004). Femtomole mixer for microsecond kinetic studies of protein folding, *Analytical Chemistry*, **75:24**, 71697178.

[7] Gad El Hak, M. (2002). *The MEMS Handbook*, Handbook Series for Mechanical Engineering, CRC Press, New York.

[8] Ivorra, B. Hertzog, D. Mohammadi, B. and Santiago, J. (2006). Global optimization for the design of fast microfluidic protein folding devices, *Int. J. Num. Meth. Eng.* **26(6)**, 319333.

[9] Lin, H. Storey, B. Oddy, M. Chen, C.-H. and Santiago, J. (2004). Instability of electrokinetic microchannel flows with conductivity gradients, *Phys. Fluids*, **16(6)**, 1922-1935.

[10] Mohammadi, B. and Santiago, J. (2001). Simulation and design of extraction and separation fluidic devices, *M2AN*, **35(3)**, 513523.

[11] Molho, J. Herr, A. Santiago, J. Mohammadi, B. Kenny, T. Brennen, R. and Gordon, G. (2001). Optimization of turn geometries for on-chip electrophoresis, *Analytical Chemestry*, **73(6)**, 13501360.

[12] Oddy, M.H. and Santiago, J. (2005). Multiple-species model for electrokinetic instability, *Phys. Fluids*, **17(1)**, 1922-1935.

[13] Probstein, R. F. (1995).*Physicochemical Hydrodynamics*, John Wiley, New York.
[14] Squires, T.M. and Quake, S. R. (2005). Instability of electrokinetic microchannel flows with conductivity gradients, *Rev. Modern Physics*, **77(3)**, 977-986.
[15] Tuomela, J. and Mohammadi, B. (2005). Simplifying numerical solution of constrained PDE systems through involutive completion, *M2AN*, **39(5)**, 909-929.

13
TOPOLOGICAL OPTIMIZATION FOR FLUIDS

13.1 Introduction

The concept of topological optimization was discovered in connection with the design of the most robust shape with an elastic material of given volume:

$$\max_{\{\Omega:|\Omega|=1\}} J(u) \; : \; -\mu\Delta u - \lambda\nabla(\nabla \cdot u) = 0, \; u|_{\Gamma_d} = 0, \; \frac{\partial u}{\partial \nu}|_{\Gamma_n} = g \quad (13.1)$$

where ν is the co-normal: $\frac{\partial u}{\partial \nu} = \mu\nabla u \cdot n + \lambda n \nabla \cdot u$. Here u is the displacement; the shape is clamped on Γ_d and subject to a load g on Γ_n. Unconstrained boundaries correspond to points where g is zero.

The problem could be solved with $J = \int_{\Gamma_n} g \cdot u$, the compliance, or equivalently the energy of the system. Let u_Ω be the solution of the elasticity equations for a shape Ω.

It has been shown [10, 7] that sequences of shape Ω^n yielding

$$J(u_{\Omega^n}) \to \sup_{\{\Omega:|\Omega|=1\}} J(u_\Omega)$$

do not converge to a set: an increasing number of tinier holes in the material always improves the design. Furthermore there is a limit but in a larger class namely, for some A:

$$\max_{\{A:f(A,\mu,\lambda)=1\}} \{J(u) \; : \; -\nabla \cdot (A\nabla u) = 0, \; u|_{\Gamma_d} = 0, \; n \cdot A\nabla u|_{\Gamma_n} = g\} \; (13.2)$$

where $x \to A(x) \in \mathbf{R}^{2\times 2}$ is the elasticity constituent matrix of a non-isotropic material which is, in general, the limit of a composite material as the laminated structure becomes finer.

A number of studies [1] have been devoted to the regularization of (13.1) such that the solution exists. In 2D, for instance, working with $J(u) + \epsilon|\partial\Omega|^2$ – i.e. a penalization of the perimeter of the boundary – insures the existence of a solution u^ϵ and convergence of u^ϵ to u solution of (13.2) when $\epsilon \to 0$. Numerical methods to compute these regularized solutions can be found in [1, 9, 4, 5].

Later, in a seminal paper, Sokolowski et al. [13] (but it seems that Schumacher [12] was earlier) studied the convergence of the solution u^ϵ of

$$u - \Delta u = f, \; \text{in } \Omega\backslash B_\epsilon(x^0), \; u|_{\partial\Omega} = 0, \; \frac{\partial u}{\partial n}|_{\partial B_\epsilon(x^0)} = 0. \quad (13.3)$$

When $\epsilon \to 0$ the solution converges to

$$u - \Delta u = f, \text{ in } \Omega, \quad u|_{\partial\Omega} = 0. \qquad (13.4)$$

However $(u^\epsilon - u)/\epsilon$ has no limit! Surprisingly $(u^\epsilon - u)\log\epsilon$ has a limit u'_T which Sokolowski and Zuchowski [13] proposed to call the topological derivative of u with respect to a domain variation about x^0. The limit is not the same if the circular hole is replaced by an elliptical hole. Anyway it opens the possibility of a quasi-gradient shape optimization method where the removal or addition of circular holes would be based on the sign of u'_T, especially after Nazarov et al. [11] obtained the variation of a functional $J(u)$ due to the perforation of the domain at x^0.

The same idea can be applied to problems with Dirichlet conditions on the hole; then shape optimization with Dirichlet condition on the boundaries could be solved with the help of such topological derivatives. Since most fluid flows have Dirichlet conditions we will restrict this chapter to these boundary conditions.

13.2 Dirichlet conditions on a shrinking hole

Let us compare the solution u of the Dirichlet problem in Ω

$$-\Delta u = f, \quad u|_{\partial\Omega} = g \qquad (13.5)$$

and compare it with u^ϵ solution of the same problem in $\Omega\setminus B(x^0, \epsilon)$. By a translation we can reduce the analysis to the case $f = 0$.

Locally around x^0 it is reasonable to consider that $u \approx a_0 + \vec{a}\cdot\vec{x}$, so that for $\epsilon \ll 1$, we can compare the two problems ($d = 2, 3$):

$$-\Delta u = 0, \text{ in } \mathbf{R}^d \quad u|_\infty \approx a_0 + \vec{a}\cdot\vec{x}$$

$$-\Delta u^\epsilon = 0, \text{ in } \mathbf{R}^d\setminus B(x^0,\epsilon) \quad u^\epsilon|_\infty \approx a_0 + \vec{a}\cdot\vec{x}, \quad u|_{\partial B(x^0,\epsilon)} = 0 \quad (13.6)$$

13.2.1 An example in dimension 2

Consider a Dirichlet problem for the Laplace operator in a disk of radius L. In polar coordinates it is: $r \in (0, L)$, $\theta \in (0, 2\pi)$

$$\frac{1}{r^2}\frac{\partial^2 u}{\partial\theta^2} + \frac{1}{r}\frac{\partial}{\partial r}\left(r\frac{\partial u}{\partial r}\right) = 0, \quad u(L,\theta) = a_0 + a_1 L\cos\theta + a_2 L\sin\theta \quad (13.7)$$

It is easy to see that $u = a_0 + a_1 x + a_2 y$.

The same Dirichlet problem in the same disk but with a round hole $\{|x| \le \epsilon\}$ is: $r \in (0, L)$, $\theta \in (0, 2\pi)$

$$\frac{1}{r^2}\frac{\partial^2 u}{\partial\theta^2} + \frac{1}{r}\frac{\partial}{\partial r}\left(r\frac{\partial u}{\partial r}\right) = 0, \quad u(L,\theta) = a_0 + a_1 L\cos\theta + b_2 L\sin\theta,$$

$$u(\epsilon, \theta) = 0. \tag{13.8}$$

When $a_1 = a_2 = 0$ then $u = a_0 (\log \frac{r}{\epsilon})/(\log \frac{L}{\epsilon})$. When $a_0 = a_2 = 0$, then $u = v(r) a_1 L \cos \theta$ where

$$-v + r(rv')' = 0, \quad v(\epsilon) = 0, \quad v(L) = 1,$$

so $u = \frac{L^2}{r} a_1 \cos \theta (\frac{r^2 - \epsilon}{L^2 - \epsilon^2})$. Similarly for a_2, so finally

$$u^\epsilon = a_0 \frac{\log \frac{r}{\epsilon}}{\log \frac{L}{\epsilon}} + \frac{L^2}{r}(a_1 \cos \theta + a_2 \sin \theta)\left(\frac{r^2 - \epsilon^2}{L^2 - \epsilon^2}\right).$$

Notice that

$$u^\epsilon = u - \frac{a_0}{\log \epsilon} \log \frac{r}{L} + o(\epsilon) + o\left(-\frac{1}{\log \epsilon}\right)$$

Therefore

$$\lim_{\epsilon \to 0} \frac{u^\epsilon - u}{(\log \frac{1}{\epsilon})^{-1}} = a_0 \log \frac{L}{r} \tag{13.9}$$

so u has a topological derivative at 0, it is $a_0 \log \frac{L}{r}$ which, by the way, is $-\infty$ at $r = 0$. A Maple graphic of the solution is shown in Fig. 13.1

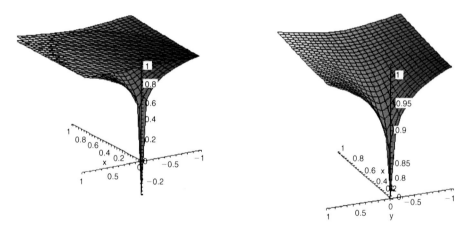

FIG. 13.1. Left: solution with a hole in the domain and comparison with the solution without (the flat surface). Right: solution by penalty with $p = 1000$ in the vicinity of $r = 0$. In both cases the part $x < 0$ is removed for clarity.

13.3 Solution by penalty

Let Γ denote the boundary of Ω, x^0 a point in Ω and let us approximate

$$-\Delta u = f, \text{ in } \Omega\backslash B(x^0,\epsilon) \quad u|_\Gamma = g, \quad u|_{\partial B(x^0,\epsilon)} = 0, \tag{13.10}$$

by

$$pI_{|x-x^0|\leq\epsilon}u - \Delta u = f, \text{ in } \Omega \quad u|_\Gamma = g, \tag{13.11}$$

where p is a positive constant destined to tend to $+\infty$. In variational form one searches for $u \in H^1(\Omega)$ with $u = g$ on Γ and

$$p\int_{B(x^0,\epsilon)} uw + \int_\Omega \nabla u \cdot \nabla w = \int_\Omega fw, \quad \forall w \in w \in H_0^1(\Omega). \tag{13.12}$$

Proposition 13.1

$$|u|_{L^2(B(x^0,\epsilon))} \leq \frac{1}{\sqrt{p}} C(|f|^2_{L^2(\Omega)} + |g|^2_{H^{\frac{1}{2}}(\Gamma)})^{\frac{1}{4}}$$

where C is a constant independent of ϵ and p. Therefore $u|_{B(x^0,\epsilon)} \to 0$ when $p \to \infty$ and so u converges to the solution of (13.11).

Proof There exists $\tilde{g} \in H^1(\Omega)$ with support outside $B(x^0,\epsilon)$ such that $\tilde{g} = g$ on Γ. Working with $v = u - \tilde{g}$ brings us to the case $g = 0$. The proposition is proved by choosing $w = u$ in (13.12).

Proposition 13.2 *Let u_p be the solution of (13.12) and u the solution of (13.5). The derivative $v = \epsilon^{-2}\lim_{\epsilon\to 0}(u_p - u)$ satisfies*

$$\int_\Omega \nabla v \nabla w = -p\bar{u}_p(x^0)w(x^0), \quad \forall w \in H_0^1(\Omega)$$

if the principal value

$$\bar{u}_p(x^0) = \lim_{\epsilon\to 0} \frac{1}{\pi\epsilon^2} \int_{|x|<\epsilon} u_p(x)\,dx$$

exists.

Proof By subtraction of (13.12) from (13.5) in variational form

$$\int_\Omega \nabla(u_p - u)\nabla w = -p\int_{B(x^0,\epsilon)} u_p w.$$

When the data are smooth there is enough regularity to approximate the integral on the right-hand side by $-pu(x^0)w(x^0)\pi\epsilon^2$.

Much is hidden in the hypothesis: *when there is enough regularity*. Rather than a mathematical investigation of the limit process we will construct an analytical solution of the problem in the next paragraph. Proposition 13.2 does not contradict (13.9). It means that p should be function of ϵ to be consistent. Even though Proposition 13.2 and (13.9) differ, they are of the same nature. Numerically the two derivatives induce the same descent direction.

Even with mesh adaptation it is very difficult to capture the singularity created by the penalty term, as shown in Fig. 13.2; to know if there is a singularity at x^0 we have to proceed analytically.

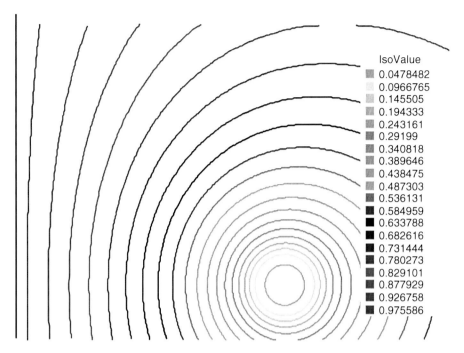

FIG. 13.2. Solution of (13.12) by the finite element method of degree 2 in the unit square when the hole is at $(0.25, 0.5)$, $\epsilon = 10^{-\frac{3}{2}}$ and $p = 10^4$.

13.3.1 A semi-analytical example

Let us again consider the case of the disk $\Omega = B(0, L) = \{|x| < L\}$ with enforcement of the Dirichlet condition on the hole $B(0, \epsilon)$ by penalty, namely solve for all $r < L$ and $\theta \in (0, 2\pi)$:

$$-pI_{r\leq \epsilon}u + \frac{1}{r^2}\frac{\partial^2 u}{\partial \theta^2} + \frac{1}{r}\frac{\partial}{\partial r}\left(r\frac{\partial u}{\partial r}\right) = 0, \qquad (13.13)$$

$$u(L, \theta) = a_0 + a_1 L\cos\theta + b_2 L \sin\theta,$$

where I_D is the characteristic function of the set D; recall that p is a constant destined to be large so as to impose $u(r, \theta) = 0$ for all $r < \epsilon$.

As before, the solution of this system is obtained in three steps. First with $a_i = \delta_{i0}$, u solves

$$-pI_{r\leq \epsilon}u + \frac{1}{r}\frac{\partial}{\partial r}(r\frac{\partial u}{\partial r}) = 0, \qquad u(L, \theta) = 1. \qquad (13.14)$$

The solution is proportional to the Bessel function of the second kind $u = \alpha K_0(r\sqrt{p})$ in the hole and $1 + \beta \log(r/L)$ when $r > \epsilon$. To match the two we need

$$1 + \beta \log\left(\frac{\epsilon}{L}\right) = \alpha K_0(\epsilon\sqrt{p}), \quad \frac{\beta}{\epsilon} = \alpha\sqrt{p}K_0'(\epsilon\sqrt{p}).$$

Hence

$$u = cI_{r\leq\epsilon}K_0(r\sqrt{p}) + I_{r>\epsilon}\left(1 + c\log\frac{r}{L}\epsilon\sqrt{p}K_0'(\epsilon\sqrt{p})\right)$$

where $c = K_0(\epsilon\sqrt{p}) - \epsilon\sqrt{p}\log\frac{\epsilon}{L}K_0'(\epsilon\sqrt{p})$. When $\epsilon \to 0$, $u \to 1$ except at $r = 0$.

Similarly when $a_i = \delta_{i1}$ (resp. $a_i = \delta_{i2}$) then $u = v(r)L\cos\theta$ (resp $v(r)L\sin\theta$) with v solution of

$$-pI_{r\leq\epsilon}u - \frac{u}{r^2} + \frac{1}{r}\frac{\partial}{\partial r}\left(r\frac{\partial u}{\partial r}\right) = 0, \quad u(L,\theta) = 1. \tag{13.15}$$

The solution in the hole involves Bessel functions of the second kind K_1: $u = \alpha K_1(r\sqrt{p})$. Outside the hole $u = \frac{r}{L} + \beta(1 - \frac{L}{r})$. Matching u and u' at $r = \epsilon$ gives

$$\alpha K_1(\epsilon\sqrt{p}) = \frac{\epsilon}{L} + \beta\left(1 - \frac{L}{\epsilon}\right), \quad \alpha\sqrt{p}K_1'(\epsilon\sqrt{p}) = \frac{1}{L} + \beta\frac{L}{\epsilon^2}.$$

One can conclude that when $\epsilon \to 0$ and p is fixed, the solution of (13.13) is not smooth but tends to $a_0 + a_1x + a_2y$ uniformly in $\Omega\setminus B(0,\rho)$. The behavior of u when $\epsilon \to 0$ and $p \to \infty$ simultaneously is another story.

13.4 Topological derivatives for fluids

After Hassine et al. [8], S. Amstutz [2] established the following result for the solution of the Navier-Stokes equations u_Ω in a domain Ω:

$$-\nu\Delta u + u\cdot\nabla u + \nabla p = 0, \quad \nabla\cdot u = 0, \quad u|_{\partial\Omega} = u_\Gamma. \tag{13.16}$$

Theorem 13.3 *Assuming that $u_\Gamma = 0$ on $\partial B(x^0, \epsilon)$, the functional $u \to J(u) \in \mathbb{R}$ has the following expansions*

$$J(u_{\Omega\setminus B(x^0,\epsilon)}) = J(u_\Omega) + 4\pi\nu\left(\log\frac{1}{\epsilon}\right)^{-1}u(x^0)\cdot v(x^0) + o\left(\left(\log\frac{1}{\epsilon}\right)^{-1}\right) \quad \text{in 2D}$$

$$J(u_{\Omega\setminus B(x^0,\epsilon)}) = J(u_\Omega) + 6\pi\nu\epsilon u(x^0)\cdot v(x^0) + o(\epsilon) \quad \text{in 3D} \tag{13.17}$$

where v is the adjoint state, solution of

$$-\nu\Delta v - u\cdot(\nabla v + \nabla v^T) + \nabla q = -J_u'(u), \quad \nabla\cdot v = 0, \quad v|_{\partial\Omega} = 0. \tag{13.18}$$

13.4.1 Application

Shape optimization problems with criteria $J(\Omega)$ for which (13.17) is available are solved by the following fixed point algorithm:

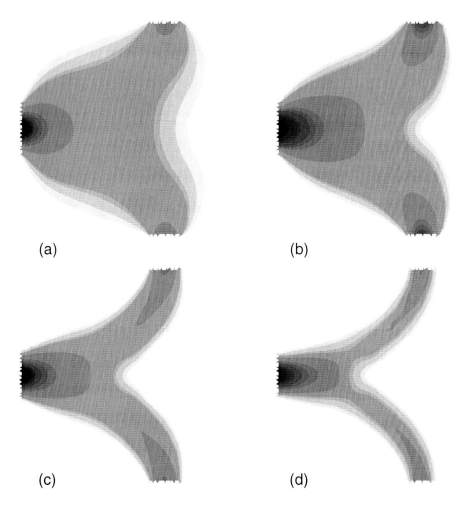

FIG. 13.3. Topological gradient for finding the best pipe with Stokes flow, four iterations of the algorithm. (Courtesy of Hassine-Masmoudi, with permission)

Topological gradient based algorithm

1. Choose a shape Ω^0
2. Loop until no change
3. Compute the flow u_Ω and the adjoint v
4. To define Ω^{n+1}
    - at all points x^0 where $u(x^0) \cdot v(x^0) > 0$ remove a small disk from the domain Ω^n
    - at all points where $u(x^0) \cdot v(x^0) < 0$ add to Ω^n a small disk around x^0.

5. End loop

The convergence of this algorithm is not known. In practice it may be necessary to moderate the changes in domain shape and take into account the connection between changes.

With a method based on the topological gradient with penalty, Hassine et al. [8] found the best pipe to transport a creeping flow (Stokes equations) given the input and output boundary. At a given flow rate one must maximize the flux through the inflow boundary. The computational domain is a square. Results are shown in Fig. 13.3 The solution was already known, in a way, because Borrvall et al. [3] studied the 3D flow in a thin channel and by assuming a parabolic profile in z, reduced the problem to a 2D flow with a varying coefficient $h(x,y)$ in the PDE, the thickness of the channel. The equations are similar to a Stokes flow with a penalty term in $h^{-1}(x,y)$, indicating that the parallel walls of the channel could collapse and block the flow.

13.5 Perspective

Topological optimization based on perforations of the domain is mathematically very interesting but numerically difficult, a situation reflected by the fact that no convergence proofs are known for $\{J(\Omega^n), \Omega^n\}$ above. On the other hand, the penalty approach allows for more general algorithms with non-constant $p(x)$ with the convention that above a threshold p_M it is a hole, below p_m it is Ω and in between it is a porous mterial. Then a standard gradient algorithm can be applied to find the optimal $p(\cdot)$ which minimizes J.

Several studies show that the level set method is a good compromise between topology optimization and boundary optimization. It was notice by Allaire et al. (see de Gournay [6]) that the level set method can remove holes but cannot create new ones, so the trick is to take an initial shape which has many holes.

References

[1] Allaire, G. (2002). *Shape Optimization by the Homogenization Method.* Applied Mathematical Sciences 146, Springer, Berlin.

[2] Amstutz, S. (2005). The topological asymptotics for the Navier-Stokes equations. *COCV ESAIM: Control, Optimisation and Calculus of Variations*, **11**, 401-425.

[3] Borrvall, T. and Petersson, J. (2003). Topological optimization of fluids in Stokes flow,*Int J. Numer. Meth. Fluids*, **41(1)**, 77-107.

[4] Bendsoe, M. and Kikuchi, N. (1988). Generating optimal topologies in structural design using a homogeneization method, *Comp. Meth. Appl. Mech. Eng.*, **71**, 197-227.

[5] Garreau, S. Guillaume, P. and Masmoudi, M. (2001). The topological asymptotic for pde systems : the elasticity case, *SIAM J. Control Optim.*, **39(6)**, 1756-1778.

[6] de Gournay, F. (2006). Velocity extension for the level-set method and multiple eigenvalues in shape optimization, *SIAM J. Control Optim.*, **45(1)**, 343-367.
[7] Grabovsky, Y. and Kohn, R. V. (1995). Topological optimization, *J. Mech. Phys. Solids*, **43(6)**, 949-972.
[8] Hassine, M. and Masmoudi, M. (2004). The topological asymptotic expansion for the quasi-Stokes problem, *ESAIM: COCV*, **10**, 478-504.
[9] Kikuchi, N. and Bendsoe, M. (2004). Topology Optimization, Springer, Berlin.
[10] Murat, F. and Tartar, L. (1997). H-convergence. *Topics in the Mathematical Modelling of Composite Materials*, A. Cherkaev and R.V. Kohn eds., series: Progress in Nonlinear Differential Equations and their Applications, Birkhauuser, Boston 1997. French version: mimeographed notes, séminaire d'Analyse Fonctionnelle et Numérique de l'Université d'Algier (1978).
[11] Nazarov, S. and Sokolowski, J. (2002). Asymptotic analysis of shape functionals, *INRIA report*, **4633**.
[12] Schumacher, A. (1995). *Topologieoptimierung von Bauteilstrukturen unter Verwendung von Lochpositionierungskriterien*, Universitat Gesamthochschule Siegen.
[13] Sokolowski, J. and Zochowski, A. (1999). On the topological derivative in shape optimization, *SIAM J. Control Optim.*, **37**, 1241-1272.

14
CONCLUSIONS AND PROSPECTIVES

Can we say something about the future ? Throughtout this book we have tried to describe our experience of different aspects of shape optimization for various applications involving CFD. In particular, we insisted on choices where we had to compromise between complexity and accuracy for the optimization platform.

Optimal shape design is not only a field of control and optimization, it also poses complex implementation problems to which the practical aspects require great attention. The capture of these aspects implies having good numerical talents, therefore calling for multi-skilled teams. In addition, very large computer CAD and simulation codes require rigor and efficiency for the platform to be efficient.

We think that these problems are more in the range of applications of gradient-based methods rather than genetic algorithms. Practical problems have a large number of degrees of freedom, and most cost functions seem to be smooth and differentiable with respect to the unknowns, so the only problem remaining is that there are many local minima; a global search strategy would be welcome but perhaps not at the cost of removing the speed-up that differential optimization methods give.

Topological optimization is also not necessary for flow problems. There does not seem to be radical changes of configurations which could not be foreseen at the parametrization stage.

As to the practical aspects for shape design problems, the treatment of the state equations in the vicinity of the shape is very important. We think that wall functions are essential for this purpose for turbulent flows, and not only for flows, but for any situation where multiple scales and anisotropy might be present for the state variables.

In all cases, it will be more and more necessary to consider multi physics applications with CFD being only a small part of the global problem. Adaptive coupling algorithms are therefore needed, such as those presented here, allowing the minimization algorithms, as a part of a simulation platform, to treat very complex state equations. But these may lead to much complexity; the optimization algorithm may turn out to be insufficiently robust.

From the minimization point of view, we think that future challenges will be more in the area of multi-point, multi-model, and multi-criteria optimization. Game theory and the definition of collaborative strategies in design can be a remedy for incompatible requirements during design.

Unstructured mesh adaptation will be increasingly necessary in industry, as it gives real freedom for complex geometries. But other methods like the fictitious domain method, level sets and non-conforming grids will compete in the future.

Despite all the progress of automatic differentiation of computer programs, sensitivity analysis and computation of derivatives will remain a major difficulty in OSD. Black-box simulation software will give a good future to finite differences unless the vendors include AD in their products. As models become more and more complex, people will develop their own computer programs less and less and this will reduce the applicability of traditional approaches with adjoint equations and state linearization, both requiring the source code. But perhaps some commercial software will use operator overloading, and encapsulation of automatic differentiation in direct modes will be easy to add. The reverse mode of AD, however, probably will be for the specialist as the complexity of reduced storage is not easy to design.

In the context of the above statements and except for genetic algorithms where hybrid methods are desired, we think that progress in OSD will be more on efficient and robust strategies in design for multi-disciplinary, multi-point and multi-criteria applications and not so much on innovative algorithms. This is due to the maturity of scientific computing and to the fact that scientists and engineers will consider sensitivity analysis and system optimization on a par with numerical simulation.

INDEX

$k - \varepsilon$ model, 46, 65

Active control, 227
Adaptive, 133
Adiabatic walls, 53
Adjoint, 167
Adjoint method, 81, 84
Adjoint mode, 87
Adjoint variable, 103
Advection-diffusion eq., 169
Aerodynamical coefficients, 212
Aeroelastic instabilities, 244
ALE, 131, 227
Anisotropy, 73
Armijo rule, 20, 184
Auto-adaptative control, 227
Automatic differentiation, 61, 81, 85, 165

Bernoulli, 45
Bernoulli law, 10
BFGS, 141, 175
Blade cascade, 212, 219
Boundary conditions, 13, 50
Boundary integral, 168
Boundary layer, 74
Breakwater, 15
Buffer region, 53
Buffeting, 229
By-section, 131

CAD, 116
CAD-free, 116
Calculus of variation, 21
CFL, 61, 66
Channel flow, 169
Class library, 88
Closure hypothesis, 47
Complex variable method, 83, 174
Complexity, 165
Compressibility effects, 54
Compressible flows, 48
Cone property, 18
Conjugate gradient, 141
Conservation of energy, 42
Conservation of momentum, 41
Conservation variables, 48
Consistent approximations, 187

Constraints, 9, 21, 212
Continuity equation, 41
Control space, 116
Convective operator, 50
Convexification, 98
Coupling algorithm, 236, 261
Crocco law, 56
Cruise condition, 132
Curvature, 10, 253

Damping matrices, 136, 241
Data base, 117
Data interpolation, 154
Data mining, 148
De Rham theorem, 44
Delaunay, 72, 200
Dependency, 106
Descent step size, 141
Direct mode, 107
Discrete derivatives, 33
Discretization, 62
Divergence free, 69
Domain deformation, 128
Drag coefficient, 212
Drag optimization, 8
Drag reduction, 213
Dynamic system, 144, 227

Eddy viscosity, 49
Efficiency, 165, 219
Eigen-modes, 227
Eigenvalues, 73, 241
Eigenvectors, 73
Elastic model, 136
Electroneutrality, 259
Electroosmosis, 246
Electrophoresis, 246
Elliptic systems, 129
Ellipticity, 13
Entropy, 43
Epi-convergence, 187
Ergodicity, 47
Euler equations, 9, 43
Explicit coupling, 236
Explicit domain deformation, 128
Extraction, 249

Fictitious domain method, 122

Fictitious walls, 57
Finite differences, 33, 83
Finite volume Galerkin, 61
Fixed point algorithm, 75
Flow control, 227
Fluid structure interaction, 227, 230
Flux difference splitting, 61
Flux vector splitting, 61
FORM, 82
Friction velocity, 52

Genetic algorithms, 150
Geometrical constraints, 131
Gradient, 81
Gradient methods, 19
Griewank function, 145

Hamilton-Jacobi equation, 122
Heavy ball, 141
Helmholtz equation, 12
High-Reynolds regions, 52
Hybrid Upwind Scheme, 62
Hyperbolic, 49

Ideal fluid, 43
Immersed boundary, 116
Immersed boundary method, 122
Incidence, 133
Incomplete sensitivity, 164, 168, 228
Incompressible, 44
Incompressible flow solver, 68
Inflow boundaries, 67
Inflow conditions, 50
Initial conditions, 50
Injection velocity, 229
Inlining, 91
Inter-procedural differentiation, 95, 167
Interior point algorithm, 133
Inviscid, 9, 196
Involutive, 260
Irrotational, 9
Isentropic, 44
Isothermal walls, 53

Jacobi loops, 121
Jacobian, 85

Kinematic viscosity, 44
Kolmogorov, 47
Kriging, 82, 154
Krylov subspace, 93
Kuhn-Tucker, 22

Lagrangian, 22, 85
Lamé coefficients, 8
Law of state, 42

Leading edge, 131, 216
Learning, 152
LES, 61, 228
Level set, 116, 122
Lift coefficient, 132, 212
Linearization, 84
Linearized Euler, 103
Local time step, 66
Log-law, 47
Low-complexity model, 249, 253
Low-Reynolds regions, 52

Magnetic polarization, 13
Mass matrices, 136, 241
Maxwell's equations, 12
Mesh adaptation, 103
Mesh deformation, 116
Mesh motion, 233
Method of moments, 82
Metric, 73
Metric intersection, 77
Microchannel, 259
Microfluidics, 246
Minimum weight, 7
Mixing length formula, 57
Mobility, 259
Modeler, 119
Moment coefficient, 221
Monte Carlo simulations, 82
Multi-criteria problems, 174
Multi-disciplinary, 229
Multi-element airfoil, 213
Multi-level construction, 172
Multi-objective optimization, 148
Multi-point, 10
Multi-scale, 75
Multicriteria problems, 101
MUSCL, 62, 64

Nanometric, 246
Nash equilibrium, 148
Navier-Stokes equations, 9, 27, 44
Near-wall regions, 52
Net charge density, 259
Newmark scheme, 234
Newton law, 41, 170
Newton method, 20, 141
Newtonian fluid, 41, 42
Noise reduction, 221
Nozzle, 15

One shot method, 27
Operator overloading, 88
Optimal control, 6
Optimal transport, 157
Oscillations, 120

OSD, 6
Osher scheme, 61
Outflow boundaries, 67
Outflow conditions, 50
Overparametric, 192

Parameterization, 116
Parametric, 117
Pareto equilibrium, 148
Pareto front, 148
Passive control, 227
Penalty, 21
Periodic, 219
Plan-form, 131
Positivity, 64
Potential, 44
Principal Component Analysis, 148
Projection scheme, 69
Protein folding, 255
PSI scheme, 70

RANS, 61, 228
Redefinition of cost function, 174
Reduced order models, 82, 173
Reduced-order models, 152
Regularization, 191
Reichardt law, 53
Resonance, 227
Response surface, 82
Reverse mode, 87, 107
Reynolds equations, 46
Reynolds hypothesis, 47
Reynolds number, 46
Reynolds relation, 58
Riblets, 11
Rigid body, 229
Robustness, 241
Roe scheme, 61
Runge Kutta scheme, 65

Self Organized Map, 148
Sensitivity, 81
Separation, 54, 249
Shape deformation, 116
Shape smoothing, 120
Shock, 43
Skew, 251
Slip boundary, 50
Sobolev space, 28
Sonic boom, 215
SORM, 82

Source code, 119
Source terms, 70
Sparse grids, 156
Spline, 192
Stability, 241
Stacking, 247
State constraints, 21
State equation, 9, 84
Steady, 131, 167
Stealth wings, 11
Steepest descent, 19, 141, 184
Stegger-Warming scheme, 61
Stiffness matrices, 136, 241
Stokes formula, 41
Stokes flow, 35
Storage, 166
Stream function, 14
Streamline, 44
Structural characteristics, 240
Supersonic, 213
SUPG, 62
Sutherland law, 48
Symmetry boundary condition, 67
Synthetic jets, 11

Tapenade, 106
Thermal conductivity, 42
Thickness, 10
Time-dependent, 11
Time-independent shapes, 11
Topological optimization, 17
Trailing edge, 10, 131
Transition, 227
Transonic, 11, 46, 213
Transpiration conditions, 11, 67, 130, 168, 188, 227
Triangulation, 62, 72

Unsteadiness, 227
Unsteady, 176
Unsteady flows, 54
Unstructured meshes, 57

Van Albada, 62
Variational form, 28
Voronoi, 207

Wall functions, 48, 130
Wall laws, 51, 66, 67, 130
Wasserstein distance, 157